Principles of Experimental Design for the Life Sciences

Murray R. Selwyn

Statistics Unlimited, Inc.
Wellesley Hills, Massachusetts

CRC Press
Boca Raton   New York   London   Tokyo

Publisher: Bob Stern
Marketing Manager: Becky McEldowney
Designer: Denise Craig
Project Editor: Gail Renard
Manufacturing Assistant: Sheri Schwartz

Library of Congress Cataloging-in-Publication Data

Selwyn, Murray R.
    Principles of experimental design for the life sciences / Murray
R. Selwyn.
        p.   cm.
    Includes bibliographical references and index.
    ISBN 0-8493-9461-9 (alk. paper)
    1. Medicine—Research—Statistical Methods. 2. Experimental design. I. Title.
R853.S7S46 1996
610′.72—dc20
                                                                96-321
                                                                CIP

No claim to original U.S. Government works
International Standard Book Number 0-8493-9461-9
Library of Congress Card Number 96-321
Printed in the United States of America 1 2 3 4 5 6 7 8 9 0
Printed on acid-free paper

The Author

Murray R. Selwyn, Ph.D., earned his bachelor's degree in mathematics and master's degree in applied mathematics from the Polytechnic Institute of Brooklyn and his Ph.D. in mathematical statistics from the George Washington University. He is currently president of Statistics Unlimited, Inc., a biostatistical consulting company in Wellesley Hills, Massachusetts, and adjunct associate professor in the mathematics department at Boston University. He has held a number of prior positions, including Director of Research Statistics and Data Systems at CIBA-GEIGY Pharmaceuticals and Mathematical Statistician at the U.S. Food and Drug Administration. Dr. Selwyn's research interests include statistical methods for biostatistics, statistical computing, and Bayesian methodology.

Dr. Selwyn is a member of the American Statistical Association, the Biometric Society, the Drug Information Association, the Society for Risk Analysis, and the Society of Toxicology. Dr. Selwyn has published widely on methods in statistics and biostatistics and on collaborative research with biomedical colleagues.

Preface

Many researchers have a hard time with statistics. They are put off by the detailed and complex mathematical treatment in most introductory courses and textbooks. As a result, they often fail to apply statistical principles of experimental design to their own studies. Such studies may therefore be poorly designed and frequently are unable to achieve their purported objectives.

Principles of Experimental Design for the Life Sciences is intended for clinicians, biological scientists, and engineers involved in the design, conduct, and interpretation of biomedical studies. By presenting experimental design topics in an understandable and nontechnical manner, free of statistical jargon and formulas, I hope to acquaint readers with the fundamental design concepts in a way that makes them easy to understand and apply in practice.

The text is replete with real-life examples taken from my 25-year career as a biostatistical consultant. These examples illustrate the main concepts of experimental design and cover a broad range of application areas in both clinical and nonclinical research.

Murray R. Selwyn
Wellesley Hills, Massachusetts

To Cecile,

for 25 years of love and devotion

Acknowledgments

The author would like to thank P. P. Charming, Thomas Copenhaver, Roy Tamura, and Eric Wakshull for their insightful and constructive comments on a previous version of the manuscript.

TABLE OF CONTENTS

Chapter 1 Introduction and Overview 1

Chapter 2 Planning Biomedical Studies 9
  2.1 Study Objectives 9
  2.2 The Planning Process 11
  2.3 Writing the Protocol 12
  2.4 Correspondence Between Objectives, Design, and
       Analysis 13

Chapter 3 Principles of Statistical Design 17
  3.1 Bias and Variability 17
  3.2 Identifying and Quantifying Sources of Bias and
       Variability 18
  3.3 Methods to Control Bias and Variability 22
  3.4 Defining the Experimental Unit 26
  3.5 Randomization 27
  3.6 Uniformity Trials 30
  3.7 Blocking 32
  3.8 Binding 34

Chapter 4 Sample Size Estimation 37
  4.1 Statistical Context 37
  4.2 Sample Sizes for Point Estimation 39
  4.3 Sample Sizes for Interval Estimation 40
  4.4 Sample Sizes for Hypothesis Testing 41
  4.5 Pilot Studies 45
  4.6 Subsampling Issues 48
  4.7 Sensitivity Analyses 49

Chapter 5 Common Designs in Biological Experimentation 53
  5.1 The Completely Randomized Design 54

5.1.1 The One-Way Layout..55
5.1.2 Factorial Design..55
5.1.3 Advantages and Disadvantages of the Completely
       Randomized Design..58
5.2 Stratified Design/Randomized Block Design 58
5.3 Crossover Study 62
5.4 Split Plot Design..65
   5.4.1 Origin 65
   5.4.2 Repeated Measures Design 66
   5.4.3 Summary..67
5.5 Types of Control 67
   5.5.1 Negative Control 68
       5.5.1.1 Subject as His/Her Own Control................68
       5.5.1.2 Untreated Control 69
       5.5.1.3 Placebo Control...69
   5.5.2 Active Control..70
5.6 Dose Selection in Dose-Response Studies.............................72
5.7 Multicenter Studies..78
5.8 Summary...80

Chapter 6 Sequential Clinical Trials..................................83
6.1 History 83
6.2 Rationale 85
6.3 Sequential Designs..86
6.4 Group Sequential Designs 93
6.5 Interim Analyses..99
6.6 Data Monitoring Boards 99

Chapter 7 High Dimensional Designs and
Process Optimization...103
7.1 Fractional Factorial Design..103
7.2 Response Surface Methodology..110
7.3 Process Optimization...114
   7.3.1 Sequential Simplex Design 114
   7.3.2 The Method of Steepest Ascent 116
   7.3.3 Experiments with Mixtures 116

Chapter 8 The Correspondence Between Objectives, Design,
and Analysis — Revisited...119
8.1 Data Analyses vs. Study Objectives and Design.................119
8.2 Types of Data 120
8.3 Verification of Assumptions 121

8.4   Multiplicity Adjustments 123
8.5   Statistical Packages 124
8.6   Analysis Strategies 126
       8.6.1   Parametric Methods for Continuous Data 127
       8.6.2   Nonparametric Methods for Continuous Data 130
       8.6.3   Methods for Discrete and Categorical Data 133
8.7   Meta Analyses... 135

Chapter 9 Summary and Concluding Remarks............................ 137
9.1   The Role of the Statistician.. 137
9.2   Summary.. 139
9.3   Concluding Remarks 140

References... 143

Appendix A. Glossary of Statistical Terms 147

Appendix B. Formulas for Sample Size Estimation......................... 151
B.1   Sample Size Formulas for Point Estimation......................... 151
B.2   Sample Size Formulas for Interval Estimation 152
B.3   Sample Size Formulas for Hypothesis Testing.................... 153

Index 157

chapter one

Introduction and Overview

It is hard to argue with anyone who advocates planning. Yet many biomedical studies are conducted with little or no thought given to the statistical aspects of the experimental study design. As a consequence, these studies are often seriously flawed and are incapable of meeting the purported purposes of the studies or are designed so that the number of experimental units is too small to have much of a chance of achieving these objectives. On the other hand, some studies are designed with far too many experimental subjects or sampling units of various kinds, so that valuable resources of the research organization (time, subjects, money) are wasted.

Why is this? I believe that there are three fundamental reasons for this phenomenon. First, the primary designers of these studies — clinicians, biological researchers, and engineers — have limited training in statistics and the application of statistical concepts in the planning of experiments. Second, these individuals may not have ready access to a consulting statistician, either within or external to their organizations, who could assist them regarding potential experimental designs. Third, even if the scientists have some statistical training themselves, introductory statistical courses largely focus on data analysis, formulas, and computations, rather than on experimental design principles. Even many statistical texts, which claim to be about design, spend most of their efforts discussing analysis of results once certain designs are selected. It is easy for individuals without a strong mathematical basis to want to avoid such drudgery.

The purpose of this book is to present topics in the area of experimental design for biomedical studies. The material is treated at the conceptual rather than at the technical level. I believe that it is much more important for biomedical researchers to understand the concepts involved in designing good studies, rather than in the nitty gritty of their applications. There will be no mathematical formulas presented in this (statistics) book (at least until Appendix B)! Instead, we will provide appropriate references for each of the topics covered, so that the reader who wants to pursue a given topic can do so by investigating the reference material or related works. This is not to say that detail and application are unimportant. On the contrary, it is not possible to completely design a satisfactory study without pursuing some of this detail. However, it seems to me that it is fundamental that concepts be well understood before the details can be decided upon.

Several examples may help to motivate some of the ideas which are discussed in subsequent chapters. These examples are mentioned briefly here, but are presented in more detail later. Their purpose at this juncture is to illustrate how the inclusion of statistical thinking and concepts at the planning stages of studies can help to ensure the studies' effectiveness and reliability.

Example 1.1

The first example is concerned with the design of studies to assess the sedative/excitability aspects of potential new drug compounds in a rodent model. After performing several such experiments, a researcher noticed considerable variability in results depending on the location of the animal cage in a rack of 24 cages with six cages per row and four rows. A uniformity study (a study without differing treatment conditions) was performed to identify and quantify sources of variation. The study highlighted the fundamental problem, and the laboratory conditions were subsequently corrected in order to make cage location an insignificant factor in future studies.

Example 1.2

In a study to examine the effect of a test compound on hepatocyte (liver cell) diameters, the experimenter decided to study eight rats per treatment group, three different lobes of each rat's liver, five fields per lobe, and approximately 1,000 to 2,000 cells per field. Most of this process was manual in that the diameter of each cell was ascertained by examining it microscopically and recording the estimated diameter. This process was quite tedious as the total amount of laboratory work was proportional to the total number of cells measured in the study. The investigator complained that because of tight deadlines and the overwhelming amount of work involved in the study, he had set up a cot in his office/laboratory and was spending 24 hours a day in the lab in order to complete the study. A sample size evaluation conducted after the study was completed (unfortunately) indicated that sampling as few as 100 cells per lobe would have been satisfactory. Thus, the study could have been accomplished with approximately one tenth of the total number of liver cells without appreciable loss of information.

Example 1.3

In a clinical study of a temporary oxygen carrying blood replacement, an initial sample size calculation indicated that approximately 90 patients were required to demonstrate superiority over control (no treatment). Because recruitment of suitable patients was expected to be difficult and determination of success or failure per case was within the first 24 hours, a sequential design was utilized. In order to collect sufficient safety data on the product, it was decided to postpone the first interim analysis of the data until approximately 60 patients had been completed. Subsequent analyses were to be performed after every 10 patients. An interim analysis conducted at this stage clearly indicated the efficacy of the product. In fact, if the planning committee for the study had not specified a minimum N of 60, it is possible that such a conclusion could have been reached even earlier. The reason that this study and its sequential design were so successful was that the assumptions made in the original sample size calculation had been too cautious, in that the actual success rate for control therapy was much lower than predicted and the treatment effect was larger than expected. By utilizing a sequential design, the number of patients in the study was substantially reduced (60 vs. 90), thereby reducing both the time and expense necessary to complete the study.

What these three examples and others throughout the text have in common is that the utilization of statistical concepts at the planning stages of a biomedical study actually or potentially resulted in a substantial improvement in the study's effectiveness and ability to achieve the purported objectives. There are many researchers who are unwilling or unable to collaborate with their statistical colleagues on study design. Often, the statistician is consulted at the end of a study and asked to perform a suitable data analysis. Unfortunately, it may be too late at that time to salvage information from the study.

To me, it is much more important to plan a study carefully and consider experimental design alternatives than it is to perform a sophisticated data analysis. Suppose, for example, that an incorrect data analysis has been performed. It is almost always possible to rerun an analysis, substituting appropriate statistical methods for those that were incorrectly applied. On the other hand, if a study has been designed with major flaws then it may not be possible to correct this situation, except perhaps by starting over. Clearly, the latter would be a most undesirable situation. The research organization will already have devoted a considerable amount of resources to the study (time, money, experimental subjects, laboratory resources that could have been used elsewhere, etc.). It is not always clear that such studies should be repeated.

The major purpose of this book is to identify and discuss topics in the statistical design of biomedical experiments from a nonmathematical perspective. The goal is to motivate the designers of such studies to consider each of these issues as they plan future studies. Some may think that this approach is overly cautious and conservative, but the rewards from careful planning and consideration of every aspect of the design will far outweigh issues of immediate deadlines and pressure to complete. In the long run, those who seriously consider and deal with these issues will be far ahead of those who ignore them or treat them lightly.

In Chapter 2, we discuss the planning process for biomedical studies. I find that it is very important to articulate the precise objectives of a study. These should be as specific as possible, and they should be reasonable and attainable. The number of objectives should be limited so that the study is not compromised by trying to achieve too much in a single study. The steps in the planning process to achieve a suitable study design should all be carried out without exception. Of critical importance are considerations to minimize study bias and variation, selection of an optimal experimental design among those suitable for the study, selection of an appropriate sample size, and randomization. A written protocol should always be prepared. The protocol should provide the specific objectives of the study, its background, inclusion and exclusion criteria for experimental units, a justification of the sample size in the study, a description of the randomization process to be utilized, and a description of all study methods including the statistical methods to be used for data analysis. For many researchers, it is unclear how one selects appropriate methods for data analysis. In fact, there is a one-to-one-to-one correspondence between study objectives, the study design, and suitable methods of data analysis. We shall revisit this issue briefly in Chapter 2 and in more detail in Chapter 8.

In Chapter 3, we explore many of the fundamental principles of experimental designs in biomedical studies. In designing most studies, it is critical to eliminate or at least minimize bias and variability. Bias is a systematic difference between a measured quantity and the true value. Variability or *random variation* is the random fluctuation of measurements about a true value, but without a systematic direction. Of these two, it is more important to minimize the effects of bias, as potentially incorrect or misleading results may be achieved when substantial bias is present. Although it is also important to minimize study variability, increased variability will only lead to experiments which must be somewhat larger (i.e., are less efficient) than those conducted under optimal conditions (i.e., minimum variability). We discuss methods to identify, quantify, and control sources of bias and variability. One particularly important issue in the design and analysis of clinical trials is the effect of patient dropouts in these studies. This issue is specifically addressed in Chapter 3. In every study, the experimental unit must be identified. Typically, this is defined as the smallest unit to which treatment can be applied. The identification of the experimental unit will therefore differ from study to study. The critical role of randomization is discussed in Chapter 3. Although many experimenters downplay the importance of study randomization, I have come to believe that it is a critical element in proper study design, a fact which I did not appreciate early in my consulting career. One technique which is often useful in identifying and quantifying sources of both bias and variability is the use of uniformity trials, i.e., studies which are conducted without differing treatment conditions and whose sole purpose is to help in estimating the effects of other underlying factors. In addition to randomization, we discuss the use of other methods to control bias and variability, specifically, methods of blocking and blinding. By a block, we mean a set of units which are expected to respond similarly as a result of treatment. The idea is that if different treatments are applied to experimental units within the same block, the estimates of treatment effects will be more precise because of the similarity of units within the block. Study blinding denotes a condition under which individuals are uninformed as to treatment group, therefore blinded. By having certain individuals blinded in this manner, one potential source of bias may be eliminated from the study.

Sample size estimation is discussed in Chapter 4. We indicate how methods and formulas for sample sizes depend on the context of inference: point estimation, interval estimation, or hypothesis testing. We describe how the formulas for sample sizes arise and what considerations are involved in such determinations. We also provide references which will allow the experimenter to obtain valid sample sizes in a variety of experimental settings. The use of pilot studies, both to obtain information on variability in a future study and to help refine experimental methods, is discussed. My experience is that pilot studies are often quite helpful and I would recommend their routine use prior to a large-scale, important study. We present the issue of subsampling in Chapter 4. Often, we need to deal with several levels of nested sampling, called subsampling. We illustrated above the issue of subsampling in the context of the hepatocyte diameter study. Here, as in many studies, the issue of relative sample sizes for primary sampling units, secondary units, etc., can only be predicted based on a pilot study or extremely representative historical data.

In Chapter 5, we explore some of the most common designs used in biomedical research. Perhaps the most common design is the completely randomized design, in which each experimental unit is randomized to a single experimental condition. Another common design is where each subject is used as his/her own control, so that the comparison of the post-treatment response with the pre-treatment response is used as a measure of the treatment effect. Although this design is very common in the medical literature, it also makes some very strong assumptions in order for a correct inference to be drawn from the study. Randomized block designs are also very prevalent and have a distinct advantage in the presence of strong block effects. Crossover studies are also very popular in biomedical research. For example, the two-period changeover is commonly used for studies to assess the bioequivalence of different formulations of a drug product. Alternative designs, such as split plot designs, are less commonly used, but the statistical techniques used to analyze such studies are appropriate to situations in which repeated measures on the same individual or experimental unit are taken (longitudinal studies). In Chapter 5, we compare and contrast the advantages and disadvantages of each of these designs, so that an experimenter will be in a position to make a more informed choice of design when the opportunity arises to choose among them.

Chapter 6 presents information on sequential designs and their application to clinical trials. Sequential designs are particularly attractive when patient recruitment is difficult and/or slow and when outcomes are obtained quickly. Sequential designs have the advantage over fixed sample designs in that they provide for interim analyses and the possibility of early study termination when results are decisive. We provide information on both sequential and group sequential designs. Group sequential designs imply that a fixed number of interim analyses will be conducted during the study at particular intervals. These must be specified in advance and should be indicated in the protocol.

The issue of process optimization is presented and discussed in Chapter 7. This topic is probably of most interest to engineers who are concerned with such problems as finding the settings of several design factors to achieve an optimal yield of a process. However, other researchers may find such techniques to be useful, especially in the early stages of a research program when a substantial number of experimental factors may be under consideration. We discuss the use of fractional factorial designs in high dimensional studies to identify those factors which are most important for a future study. In addition, we provide information on approaches in response surface methodology which can be used in process optimization.

In Chapter 8, we revisit the question of correspondence between study objectives, design, and statistical analysis. Without elaborating on the details, we provide guidance on how an appropriate statistical analysis is conducted. Basically, the appropriate analysis depends on the purpose and form of the study and the properties of the data that are being statistically analyzed.

Chapter 9 provides a summary of the experimental design process and a discussion of the critical role of the statistician. Unlike most scientists who can often work in isolation from colleagues in other fields, the statistician is dependent on his/her scientific colleagues to provide the opportunity to collaborate on projects of mutual interest. In return, he/she provides colleagues with particular expertise in design/analysis which is useful in ensuring that the best possible studies are conducted. Usually, the statistician is not an expert in the particular subject matter being studied. For this, he/she relies on the clinician, biological researcher, or engineer for expertise. What the statistician brings to the process is the training and experience obtained in a wide variety of experimental settings which allow him/her to have an overview of possible alternative experimental designs and approaches, and a feel for problems or issues which are likely to arise. For example, the statistician cannot necessarily identify all the factors that should be studied in a uniformity trial, but he/she can assist the experimenter in designing a suitable study once such factors have been identified. By utilizing the particular skills and training of each individual involved in the design and analysis of a biomedical study, such a multidisciplinary approach will result in a better and more efficient study and one which is most likely to accomplish its objectives.

In writing this book, I am attempting to convince biomedical researchers to seriously think through statistical considerations at the design stage of a study. Throughout my career, I have often seen studies which were haphazardly designed and as a result, have had disastrous consequences. Many researchers are put off by the mathematical nature of statisticians and statistics texts. These texts are usually replete with complicated formulas and mystifying symbols (i.e., "it's all Greek to me"). By discussing experimental design topics in a nontechnical way, I hope to persuade researchers of the importance of statistical planning in the design of their studies. Such planning efforts should pay off by enabling researchers to perform the best studies possible. Another potential payoff will come when the results of such studies are submitted for publication in biomedical journals. Many journals are beginning to routinely include a statistical review by a qualified statistician as part of the review process.

chapter two

Planning Biomedical Studies

2.1 Study Objectives

An initial step in the planning process for biomedical studies is the specification of study objectives. But even before one considers what the precise objectives of a study should be, one must realize how that study will be integrated into the set of related studies on the experimental compound, drug, process, or other subject of the study. How will this particular study relate to management objectives? What has been learned from previous studies? What studies are planned for the future? How will the results from this study contribute to the overall body of knowledge about the subject? In addition, one must consider how a particular study will help to satisfy regulatory requirements (e.g., assessment of safety and efficacy of a new drug product).

Once it has been decided to conduct the study, the initial step in the planning process is to specify the study objectives. Study objectives should be well defined and written as explicitly as possible. For example, it is not enough to say the objectives of the study are to evaluate the safety and efficacy of Drug X. This could define almost any study being conducted.* In the context of this example, one would need to identify what was the target population** under study (e.g., normal volunteers, a particular patient group, etc.), what were the particular measures of efficacy being examined, how was safety to be determined, and so on.

The number of objectives in the study should be limited. I believe that three objectives should be the maximum number under investigation in almost every study. In my experience, to have more usually means compromising the integrity of the study. By having too many

---

* Unlike drugs, there are no generic study objectives.
** A distinction can also be made between the study population and the target population. The former is the collection of experimental units from which the sample is selected (generally through some form of randomization).

objectives and trying to accomplish each, the study resources are stretched too thin and as a result, often none of the study objectives is satisfied and valuable resources are completely wasted.

Example 2.1

As an example of this phenomenon, I recall that early in my career, I collaborated with Dr. Richard T. on experiments he was conducting. His typical set of protocol objectives was as given in Table 2.1. Of particular concern were not only the overwhelming number of proposed objectives of the study (often ten or more), but also the fact that the successful completion of many of the primary objectives (e.g., determination and comparison of efficacy of various drugs) depended on critical assay processes (i.e., objective #11 in Table 2.1) which had not been completely validated prior to the start of the study. Suppose that the assay validation portion of the study went poorly and in fact, the acceptability of the assay was still questionable at the end of the study. In this case, the whole study would be compromised and the experimenter would have to start all over. I suppose that Dr. T. was hoping to impress (his colleagues?) with the scope of his research studies and believed that by having so many objectives, he was being efficient. On the contrary, I feel that by trying to do too much in each study, he actually sacrificed efficiency. Certainly, by performing a preliminary study in which the sole objective was to validate the assay method, he would have eliminated a possibly disastrous outcome from the main study.

Table 2.1 Sample objectives for studies conducted by Dr. Richard T.

---

1. Evaluate the dose-response relationship for Drug A.
2. Evaluate the dose-response relationship for Drug B.
3. Compare the potency of Drug A to the potency of Drug B.
   .
   .
   .

11. Validate the assay methodology used in study.

Objectives should be reasonable and attainable. One needs to be realistic in what can be accomplished in a single study. For example, if one can assess the probability of success for a proposed study (see Chapter 4 on sample size estimation) and it appears that the chances of success are very low or that too great a number of experimental units would be required to have a reasonable probability of success, then perhaps the present study should be abandoned in favor of one with a more limited scope. Or at least one should consider possible design alternatives which would make the accomplishment of the proposed objectives more realistic.

At the end of each study, one should be able to determine whether the study objectives have been met. This may seem like an obvious requirement. However, I have been involved in a number of studies in which the researcher has had a preconceived notion about how he/she expected the experiment to come out. This opinion about the outcome was so strong, in fact, that any possibility of a substantially different outcome was beyond the researcher's realm of consideration. On several occasions, such different outcomes actually occurred and as a result, no clear-cut conclusion could be made from the study. In thinking about potential study objectives, one needs to consider all possible study outcomes and decide what one might conclude from each. If there are possible study outcomes which will result in an indecisive study, then perhaps the objectives and design of the study should be rethought to eliminate such a possibility.

2.2   The Planning Process

The essential steps in the planning process for a biomedical study include the definition of study objectives, consideration of and selection among alternative experimental designs, estimation of sample size, writing of the protocol, and the generation of a randomization plan or schedule. Often, in preparing for a study, many of the measurement processes required in the study need to be specified and worked out in advance of the actual conduct of the study. For example, as indicated above, if a chemical or biological assay is to be an essential feature of the study, then the assay should be sufficiently validated prior to the main study. This can be accomplished during a previous or pilot study and the documentation from the validation effort should be made part of the study documentation package.

In clinical studies in human subjects, the obtaining of informed consent and institutional review board approval is necessary for such studies to be undertaken. Again, documentation of these approvals becomes part of the written record of the study.

Perhaps the least understood part of study planning from a statistical perspective is the idea that alternative experimental designs can be considered in most situations. Many of the common designs used in biomedical research are discussed in Chapters 5 through 7. It would be impossible to cover every possible statistical design, however, nor would it necessarily be productive to present information on obscure designs just for the sake of completeness. Clearly, investigators who believe that their purposes are not sufficiently being served by one of the more common designs discussed in this book are encouraged to pursue an even wider range of alternatives.

A simple example will illustrate what I mean by alternative designs. Suppose that a clinical study is to be performed comparing two potential forms of therapy, say A and B. A very common design (the completely randomized design) randomizes patients to either therapy A or therapy B. Usually an equal number of subjects are randomized to each treatment condition as this is the optimal use of subjects. An alternative design which could instead be utilized in this situation is the two period crossover design (see Section 5.4) in which subjects get both therapy A and therapy B. Half of the subjects are randomly assigned to therapy A first, and then, after a suitable washout period, receive therapy B. The remainder of subjects receive B first and then A.

Each design has its own advantages and disadvantages relative to the other. In particular, the two period crossover often requires fewer total subjects than the completely randomized design as comparisons of the two therapies are made within each subject rather than between subjects. On the other hand, because the crossover study requires an initial therapy period, a washout period, and a second therapy period, it will require a longer time interval for each subject to complete the study. It is often trade-offs like this that must be considered by study planners in determining the most appropriate statistical design for their study.

Certain study limitations will often dictate which choices can be considered. If the primary endpoint in the above study is survival (i.e., time to death), then this clearly eliminates a crossover study from consideration. However, even in this situation, there are design alternatives: completely randomized designs, randomized block designs, etc.

Many types of studies have "standard designs" which are typically used unless there is a very good reason to do otherwise. In performing bioequivalence studies, for example, in which the major objective is to compare the extent and rate of absorption of two formulations of the same drug entity, the two period crossover is the design of choice. However, even in the bioequivalence setting, special circumstances may invite one to consider alternative designs which may be more efficient in that particular situation.

The main message here is that alternative experimental designs should always be considered, even if some can be quickly eliminated due to practical limitations imposed on the study. Every design has relative advantages and disadvantages. Study designers need to be aware of them and weigh their pros and cons carefully, so that the best informed choice can be made.

2.3   *Writing the Protocol*

Every study should have a written protocol. Protocols need not be elegant or complicated, but they should be written down in some reasonable form.* A protocol should not be written after the study has been completed.

The protocol ideally would contain sections introducing the topic and providing a rationale for performing the current study (i.e., deal with the management objectives), specify the study objectives in sufficient detail, describe the experimental design and why it was selected, describe the experimental procedures to be followed, provide information on sample sizes and their justification, and describe the statistical and other data analyses to be performed. When a study has multiple endpoints, as most studies do, the protocol should identify which parameters measured are most important (e.g., primary efficacy parameters) and which are secondary.

In clinical trials, the case report forms (CRFs) or patient record forms (PRFs) should be a part of the protocol. Information on informed consent and institutional review board approval should also be included with the protocol.

In deciding whether to be more general or more specific in writing a particular section of the protocol, I believe that it is probably better to err on the side of being too specific rather than too general. For one thing, if the study is a multicenter study with several investigators, then it is imperative to ensure that all investigators are using the same methods. It would be disastrous to find out at the end of the study that each investigator got a different result, primarily because he/she was using a fundamentally different interpretation of the protocol.** Such recommendations apply to statistical procedures as well. It is very easy to describe statistical methods in such generality that they do not mean anything to anyone. This is neither desirable nor productive, unless one's main intention is to be obscure. Even if study methods need to be modified due to findings as the study progresses or other practical factors, such changes can be made during the study. Major changes should be documented as protocol amendments which become part of the overall study documentation.

2.4   Correspondence Between Objectives, Design, and Analysis

There is a one-to-one-to-one correspondence between the study objectives, the study design, and the statistical analysis (see Figure 2.1). By this I mean that the study objectives will indicate which of several

---

* Backs of envelopes and cocktail napkins do not count!
** In this regard, it is often helpful to hold an investigator's meeting at the start of a multicenter clinical study to motivate all the investigators and to impress upon them the need for uniformity. By having everyone together at the same time it is also possible to iron out any last minute discrepancies in procedures.

designs may be considered. Usually one will be selected as the best because of its relative practical advantages, as indicated above. Once the experimental design is specified, it will often be relatively easy to decide on an appropriate statistical analysis.*

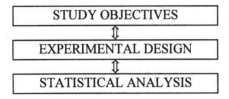

Figure 2.1 Correspondence between study objectives, design, and analysis.

Conversely, if one knows the statistical analyses to be conducted, one can infer the design of the study and its objectives. Therefore, one can go back and forth among these three because of their natural correspondence.

Nonstatisticians are often surprised how easy it is to determine appropriate statistical methods for a well-designed study. The reason is that the methods follow easily from the objectives and design. It is usually not a difficult choice and it is essentially predetermined by the protocol (irrespective of the fact that the methods themselves are often specified in the data analysis section of the protocol). Although it is often possible to consider slightly different versions of the same analysis, such differences are usually minor and will generally lead to the same basic conclusions from the study. If they do not, then this in itself would be a reason to be alerted, and fundamental features of the study would need to be investigated to evaluate the causes for such a discrepancy.

On the other hand, consider what happens when a study is poorly planned and poorly designed. Often the statistician is asked to perform a data analysis without being consulted on the study design. When the study itself has major flaws, this activity is known in the trade as a salvage job, where the statistician is merely attempting to salvage as much information as he/she can from an otherwise bad study. Sometimes, some of the information will be useful, either in decision making or at least in the planning of a future study. The former will occur if, even in the face of a poor design, the data are so overwhelming that they overcome the design flaws of the study. This rarely happens. In

* As indicated in Chapter 8, the selection of an appropriate statistical method depends not only on the design and objectives, but also on the form of the data. For example, in deciding whether normal theory procedures are appropriate, one issue is whether data are continuous or discrete (i.e., take on only a limited number of values). Clearly data which are "very" discrete should not be analyzed by normal theory methods.

the latter case, the study provides some information about study procedures and variability, so that such data can be utilized to make the subsequent (newly designed) study better. Unfortunately, in some instances the study is so badly flawed that it provides no real scientific information at all, only that the study planners failed in their efforts.

As a consulting statistician, I have found that there are two situations in which clients eagerly come to me. Either I have helped them in the past and our joint efforts have been successful, so that they want to continue in this vein, or they have had the kind of unfortunate outcomes I describe above (usually without input from me or another statistician) and they come to me after experiencing such a failure. By carefully considering and making decisions about each of the critical steps in the planning of a biomedical study, let us hope that such failures can be avoided in the future!

chapter three

Principles of Statistical Design

3.1 Bias and Variability

In large part, much of the effort in experimental design is aimed at eliminating, or at least minimizing the bias and variability from extraneous sources in a study. Although most scientists are likely to be familiar with the terms *bias* and *variability*, it is probably advisable to define them anyway, as these concepts provide much of the motivation for the foundations of study design as discussed in this chapter. By bias, we mean a systematic deviation in observed measurements from the true value. By variability, we mean a random fluctuation about a central value.

These terms are easily illustrated with a simple example. Suppose that we have an object that actually weighs 150 lbs. If we weigh the object on a scale and the average weight obtained from this process, given a large number of repeated weighings, is 153 lbs., then the scale exhibits a positive bias of +3 lbs. Suppose, on the other hand, that the average of a large number of repeated weighings is 150, but that individual weighings vary from one to another, perhaps taking the value 148 one time, 153 the next, 155 the next, and so on. Here, the measurement process exhibits variability, but no inherent bias. Of course, a process can exhibit both variability and bias, as would be the case if the scale averaged 152 over the long run and also had weights which varied from one time to another.

Other terms which are also used in this regard are *accuracy* and *precision*. If a measurement process is free of bias, it is accurate. Similarly, if a measurement process has little variability, it is called precise.

Good experiments will be as free as possible from both bias and variability. Of the two, it is perhaps more important to control bias. The reason for this is that if a study has substantial inherent bias, then one may derive an erroneous conclusion from it. On the other hand, if a study is done in such a way that major outcomes are subject to more variability than could otherwise be achieved, perhaps with greater refinement of experimental methods or other techniques, then the study may still reach the correct conclusion, but it will probably need to be larger in scope than necessary (i.e., need larger sample sizes) in order to do so.

Example 3.1

A very simple example will illustrate the effect of bias on the ability to draw correct conclusions. Suppose that a study is performed on laboratory animals and that (for some unknown reason, perhaps convenience) the experimenter allocates all males to a treated therapy and all females to control therapy. Suppose also that at the end of the study, the experimenter finds a strong difference between the treated and control groups. Can he/she conclude that a treatment effect has been demonstrated? Certainly not, as the treatment condition is completely confounded with the sexes of the experimental animals.

While it would be hard to believe that any experimenter would actually perform an experiment with such an obvious flaw, many other sources of bias (and variability) are much more subtle, but they can still create problems if left undiscovered and uncontrolled. In the remainder of this chapter, we consider mechanisms for identifying, quantifying, and minimizing the effects of potential bias and variability in designed studies. Only by carefully considering all possible sources and dealing with them in the most efficient way can an experimenter truly optimize an experimental design.

3.2   Identifying and Quantifying Sources of Bias and Variability

The first step in considering the form of an experiment is to identify the factors which might affect the experimental outcomes. In particular, one needs to consider what the target population of the study is, how a sample will be selected from the population, and the experimental processes and procedures that will be performed during the course of the study.

In specifying the target population, one must relate the sample drawn in the study to the study objectives. For example, in a clinical study of a new drug product, are potential patients to be both males and females? Is there a particular age range for which the drug is intended? For what diseases or conditions is the therapy to be applied? Clearly the subjects utilized should mimic the target population to the most reasonable extent possible. This, of course, is not to say that every clinical study needs to be performed with patients from the target population. Usually Phase I clinical studies designed to provide an early assessment of a new drug's safety profile are conducted in healthy normal volunteers.

If we are designing a study to be conducted in animals (e.g., either as a preclinical study to investigate safety and/or efficacy prior to clinical trials in humans or because the product is intended for veterinary use), we need to specify the species and strain of animal used, the age range, often a weight range, and other characteristics of the test animals that would make the study as realistic and informative as possible.

In the above discussion, we have used two terms which have special meanings in a statistical context. A *population* or *target population* is the collection of individuals or subjects about whom we wish to learn. In the context of a clinical study, for example, one population might be the set of patients who have been diagnosed as potentially having primary or metastatic tumors of the liver. The *sample* or *random sample* is a subset of the population which is actually studied in the current experiment. The term *random* here also implies that the way subjects are entered into the sample (from the population) involves some degree of randomization (see Section 3.5).* The basic idea is that the sample is representative of the population, and by performing a study with the sample, one can then make inference to the target population. However, if the sample is not representative of the population, then there is an inherent bias introduced because it is unclear how the results of the study can be applied to the population if its characteristics are different.

If we first consider the issue of bias, we see that there are many potential sources in most studies. Among these are the way in which experimental units are allocated to treatment conditions, the ways in which measurements are taken, and the association between unwanted variables** and treatment conditions or other experimental design factors of interest. It is clear that if subjects are allocated to treatment in a systematic way (e.g., all control subjects done first, or alternating control first and then treated in each pair of subjects), then this lends itself to possible bias. The primary methods for minimizing bias in allocation of subjects are randomization (Section 3.5) and blinding of

---

* In practice, it is often difficult and/or impractical to conduct certain types of studies with true random sampling from the population. In a clinical study, we generally select a small number of study sites and use patients only from those sites. Thus, only patients from these sites will constitute the study sample. However, if the study sites themselves are supposedly representative of all possible sites (for the target population under study), then the sample drawn in this way should also be representative.

** In this regard, we can define several types of variables in a study. The *outcome variables* are also called *response variables*. The study factors (e.g., treatment conditions, variables on which stratification is performed) are called *explanatory variables*. Certain uncontrollable variables may also be referred to as *nuisance variables*.

subjects and/or investigators as to the treatment condition of the subjects (Section 3.8).

In addition to the issue of who takes measurements and whether he/she is aware of the treatment condition being applied to a given subject is the question of the measurement processes themselves. One has to consider the factors affecting the measurement process and how they will ultimately influence experimental outcomes. For example, are the processes affected by season of the year perhaps due to the influences of temperature and humidity, day of the week, time of day (a.m. versus p.m.) and so on? In some cases, we know that there is an inherent circadian rhythm which can easily affect certain measurements and findings. For example, it is well known that, in people, blood pressure is typically highest around mid-morning and lowest in the early a.m. Even the measurement process itself can produce an abnormally high response (the so-called alert reaction) when measured in a physician's office by a standard sphygmomanometer.

If one of the important features of the study incorporates a biological or chemical assay, for example, then one needs to examine the properties of the assay to show that it is accurate (and precise), sensitive and specific for the target compound, and reproducible. What are the factors that determine the results of the assay? Suppose for example that two different laboratory technicians are scheduled to perform the assays in an upcoming study. Are there appreciable differences obtained by the two technicians? If an experimenter does not investigate this issue prior to the conduct of the study, and in fact, such differences do exist, then some of the effects seen in the study can in part be due to differences between technicians. This is another example of confounding, where differences between technicians, an unwanted factor, may be partially associated with apparent treatment differences.

Usually it is up to the scientific investigator to identify the various factors which may affect the measurement processes and experimental outcomes. These can then be examined in preliminary or pilot studies and the magnitudes of their effects can be quantified. Suppose that in the example given above, one finds that assays performed by one laboratory technician consistently result in higher values than those performed by the other technician. One would then investigate the causes of these differences. Perhaps their techniques are slightly different and so the cause can be attributed to some particular aspect of the assay process. If it is possible to standardize this aspect and insure that both technicians follow the standardized routine in the future, then this is the most preferable outcome. Any potential bias in the actual study will thus be eliminated or at least substantially reduced. Once such a discrepancy is found and corrected, a confirmatory study might be performed to make sure that this indeed was the cause of the different outcomes for the two technicians.

Sometimes the cause for such a discrepancy cannot be discovered, even with several preliminary studies and extensive investigation. In such cases, the experimenter may be forced to slightly alter the experimental design to take into account factors beyond his/her control. We will explore this topic in more detail in the next section. In the lab technician example, two approaches to eliminate potential biases are (1) to have all the assays for a given study performed by only one of the technicians if this is a feasible workload, or (2) to split the assays equally between the two technicians and do this in such a way that the treatment conditions are also equally balanced for each one.

One issue that is a particular source of bias for human clinical trials is the issue of study dropouts. Here, by dropouts, we mean experimental subjects who do not complete the study, and when such events are an unplanned part of the experiment. Subjects may drop out of a study for a number of reasons, some of which may be totally unrelated to the study. For example, in a long-term study, some subjects may move their places of residence and be unable to continue in the study. If the move or other similar reasons are unrelated to the disease process or treatments under study, then the dropouts are random events (not causally related to treatment), but they do reduce the effective sample size in the study. On the other hand, sometimes subjects drop out of studies because of toxicity of a drug product or the fact that it is not efficacious for them. In these cases, dropouts can potentially bias the outcomes of the study. For example, if subjects drop out due to lack of efficacy, then any estimate of the treatment's efficacy based on the remaining patients will overstate its benefits. Similarly, if dropouts are caused by toxicity, estimates of side effects using only patients completing the study will underestimate the true population rates.

In addition to studying potential sources of bias, preliminary studies should often be conducted to identify and quantify potential sources of variability. Again, it is probably most appropriate for the experimental investigator to initially identify potential sources. A consulting statistician may be helpful in identifying additional sources which have not been considered by the investigator. In general, it is probably better to over-specify the list rather than under-specify it. It may be easy to eliminate some factors as inconsequential once they have been listed.

In many studies, there is often a large number of sources of variability. For example, we often are faced with such factors as subject variability (patient to patient, or animal to animal), within subject variability (i.e., subsampling by biological unit within subjects such as cells, etc.), measurement day, day of assay, laboratory instrument used, technician, source of reagents, etc. The main issue is to identify which of all possible factors contribute most to variability. These can then be controlled in various ways as discussed below.

In planning a large or important study, it is probably advisable to perform a number of preliminary or pilot studies to assess the impact of factors on study bias and variability. Usually a series of small studies is preferable to one large pilot study. This sequential process may be considered as an upgrading or refinement of the experimental conditions under which the actual study will be performed. One option that is sometimes used in this regard is to study experimental conditions in the absence of treatment or using a reference standard only. Such a *uniformity trial* is described and illustrated in Section 3.6.

3.3   Methods to Control Bias and Variability

Once the major sources of bias and variability have been identified, there is a wide variety of methods to deal with them. As indicated above, if the specific causes of bias and / or variability can be discovered and eliminated through refinement or standardization of experimental techniques or conditions, then this is the ideal solution to the problem. In cases where it is impossible to track down a specific cause or it is infeasible to implement an experimental methodological remedy, other methods must be utilized.

In order to minimize potential bias throughout the conduct of studies, it is imperative that they have in place written protocols. If a study involves multiple investigators or experimenters, it is critical that they all "buy in" to the protocol and follow it throughout the study. In this regard, in multicenter clinical trials, an investigators' meeting scheduled just prior to the start of a study is often useful in identifying and resolving any misinterpretations and misunderstandings about the clinical protocol.

Similarly, it is a good idea to provide specific definitions of terms and / or rating levels, especially for subjective measurements. In both clinical and preclinical studies, variables are often rated on an ordinal scale (only the order matters, there is really no absolute numeric value as there would be for something like body temperature). In order to provide guidance when multiple raters are involved, as is almost always the case, standard definitions for the levels of the subjective measurement should be provided. For example, in a double-blind placebo controlled study to assess the safety and efficacy of a biological product on synkinetic closure of the eyelid associated with aberrant regeneration, the protocol provided specific definitions on a seven point scale for physicians to use. As can be seen from Table 3.1, such "anchor points" provided raters with sufficient information for all of them to provide consistent ratings. In addition, to further limit variability, the protocol specified that pre- and post-treatment assessments on the same patient were to be performed by the same physician.

Table 3.1 Physicians' grading scale for severity of synkinesis

Severity Grade	Description	Definition
0	Not present	
1	Mild	Upper lid closure of < 1.5 mm
2	Mild-Moderate	Upper lid closure to within 1 mm of corneal light reflex
3	Moderate	Upper lid moves to the position of the corneal light reflex
4	Moderate-Severe	Corneal light reflex is obliterated by upper lid movement
5	Severe	Corneal upper lid moves to position of inferior limbus at the 6 o'clock position with evidence of scleral show
6	Very severe	Eyes close completely with lower facial movement

Example 3.2

In addition, when multiple raters are to make independent assessments, it is often productive to have prestudy training sessions using reference cases to promote standardization of ratings to the greatest extent possible. For example, in Phase II/III studies of a contrast enhancing agent for magnetic resonance imaging (MRI) of the liver and spleen in patients with known or suspected liver pathology (liver tumor or other primary tumor with potential metastasis to the liver), prereading training sessions were held with radiologists who participated in the studies as "blinded" assessors. Through the use of standard radiographs obtained in earlier studies, these independent assessors were prepared to read images in the current study by alerting them as to the properties and effects of the agent. The goal was to minimize study variability of the independent assessments of images by providing each radiologist with the same examples and training. In cases where investigators or raters can be gathered together prior to the start of a study, they can discuss the particular methodological issues involved in making assessments and often resolve potential discrepancies by reaching consensus ratings on training cases.

Although many sources of potential experimental bias can be eliminated or reduced through methods such as those described above, there still may remain additional sources which are beyond the control or awareness of the study designers. Here, general principles of statistical design can be brought to bear to alleviate such potential problems. The choice of experimental design can be made to minimize the biasing effects of extraneous factors. If such factors have been identified, then they can be used as *blocking variables* and all treatment conditions can be equally represented within each block. For example, in the preclinical animal study mentioned above, where sex of the animal is a potentially important factor, males and females can be equally represented in the study and treatment conditions can be equally represented in males and equally represented in females. Suppose, for instance, that there are three treatments to be applied (A, B, and C) and that a sample size of six animals is deemed sufficient per treatment group and sex. The allocation of animals to treatment groups would be as specified in Table 3.2.

Table 3.2   Sample sizes for preclinical animal study
with both sexes

	Treatment A	Treatment B	Treatment C
Males	6	6	6
Females	6	6	6

In addition to randomized block designs (see Section 5.3), other designs might also be considered. The main idea is to use the experimental design which is most suitable for the situation at hand. Bias control is a primary concern.

It would be impossible, however, to be able to identify and resolve every potential factor that might affect an experimental outcome. While major factors can often be identified and controlled, others cannot. In this regard, a major strategy to control bias in almost every experimental study is the use of randomization to allocate experimental units to differing treatment conditions. In addition to being a scientifically acceptable method, randomization automatically balances uncontrolled variables (in the long run). The use of randomization to control study bias is further discussed in Section 3.5.

Another technique which is often useful in minimizing study bias is the blinding of investigators, subjects, and raters as to the treatment group identity of the subject. Clinical studies in which the subject is unaware of his/her treatment condition are called *single-blinded*, whereas studies in which both subjects and investigators are unaware are called *double-blinded*. We have already discussed how blinded ratings are beneficial in reducing potential study bias. The idea of blinding will be explored further in Section 3.8.

One issue which is of particular concern as it relates to potential study bias is the issue of dropouts in clinical studies. When dropouts occur for reasons which may be related to treatment (such as lack of efficacy or side effects), estimates based only on remaining patients may be biased. For this reason, it is important to minimize the dropout rate in a clinical study. When dropouts do occur, the reasons for them should be investigated. Every reasonable effort should be made to obtain information on study outcomes for each subject entered into the study for as long as it is practical to do so. Studies in which the dropout rate is excessive are open to question and are less likely to be accepted by regulatory agencies (Chiacchierini and Seidman, 1991).

There are also several major strategies to controlling variability in a study. It may be clear from preliminary or pilot studies that one or several factors are contributing substantially to the overall study variability. For example, it may be observed that a major experimental parameter varies considerably with the age of the subject. In a study in which age of the subject is not of great importance, one may therefore limit the range of allowable ages in order to reduce variability in this particular response. However, one must realize that there is a tradeoff between those variables which are held constant or allowed to vary over a narrow range, and the degree to which one will be able to make inferences to a larger population (i.e., one which allows for greater diversity) from the restricted sample.

Variability considerations will also often affect the choice of study design. For example, in some instances, responses among subjects are highly variable, whereas responses in the same subject (if it is possible to have repeated measurements) are relatively close. In such cases, crossover designs (see Section 5.4) are generally preferable to designs in which each subject receives only a single treatment, as the crossover will result in substantially fewer subjects being required than a completely randomized design.

The major strategy to overcome variability is through replication. The general idea, of course, is that estimates based on more experimental units will be more precise than estimates based on just a few units. This question of sample size is discussed in detail in Chapter 4.

The issue of an appropriate sample size is further complicated when there are multiple levels of sampling. The process of taking samples at levels below the primary level is known as subsampling. When an outcome variable can only be assessed at a subsampling level, then it is important to see how each level of sampling contributes to the overall variability.

Example 3.3 (Continuation of Example 1.2)

To illustrate this phenomenon, I will describe an actual experiment that was conducted and how it would have benefited from some small preliminary studies to assess variation. In a study to determine the effects of a particular chemical on hepatic (liver) changes in rats, the investigator was faced with multiple levels of sampling: animals, lobes of the liver, fields within lobes, and cells within fields. In the actual design utilized, ten animals were randomly allocated to each of two groups (control and treated). Five fields were microscopically examined for each of the three lobes of the liver and a total of between 300 and 2100 cells were studied per lobe for each rat. Usually the number was 1000 or more. The hepatocyte diameter was the measurement of interest. The total amount of experimental effort was proportional to the number of cells examined microscopically, approximately 50,000.

From our analysis of the components of total variability, we estimated that the majority of the variation was between animals, so that very little information was derived from sampling many cells per animal. In fact, our sample size analysis indicated that 100 cells per animal would have sufficed and provided adequate statistical power. The total number of cells could have been about 6,000, a reduction of almost 90% in comparison to what was actually done.

The moral of this story is that replication is important in reducing variability, but it must be replication at the right level. By performing preliminary studies in which the components of total variability are assessed and then by acting upon this information, one is able to design much more efficient studies.

3.4   Defining the Experimental Unit

In designing studies, it is important that the experimenter understand how the experimental unit is defined. In many studies, the definition is obvious, but especially in studies with multiple levels of sampling, care must be exercised so that the sample size is specified at the right level. In general, the experimental unit is the smallest unit to which treatments (or differing experimental conditions which are being studied) can be applied.

A number of examples will help to clarify this definition. In laboratory studies in rodents (rats or mice), in which the animals are multiply caged and the same treatment is applied to each of the animals in the same cage, the cage is the experimental unit, rather than the animal. Similarly in growth promotant, therapeutic, and prophylactic studies of veterinary products where animals are housed in pens and the same treatment is applied to all of the animals in the pen, the pen is the experimental unit, rather than each animal. In the example mentioned above, involving differing treatments administered to rats, three lobes of the liver, several fields per lobe, and many cells per field, the experimental unit was the rat. The important feature here is that although many cells were ultimately assessed for hepatocyte diameter, since the experimental unit is really the animal itself rather than any of the lower levels of sampling, it is important to adequately replicate at the animal level in order for the study to be sufficiently large to meet its intended purposes. In the context of this study, the animal is the experimental unit because the same treatment is applied to all lobes of the liver and all cells for a single animal.

In a clinical study designed to assess the skin irritation potential of five different dermatological compounds, all five compounds were administered simultaneously on the backs of normal volunteers in the study at five separate test areas labeled A through E. Because each test area was randomly assigned to one of the five different treatment conditions, it was the experimental unit in the study rather than each volunteer being an experimental unit. In a clinical study of the safety and efficacy of a polymer and delivery system for treatment of postinflammatory, postsurgery, or endometriosis-related adhesions of pelvic organs, patients with bilateral adhesions were randomized to have the experimental therapy applied to either the left or right side of the pelvis, with the standard therapy (laparoscopic adhesiolysis) applied to the contralateral side. Here again, it was the side of the pelvis, rather than the patient which was the experimental unit. An interesting feature of clinical studies such as these is that the experimental unit, as defined here, is appropriate and usually very efficient for evaluating efficacy, but the evaluation of safety of novel treatments cannot always be made as easily.

Because multiple treatments are being applied to the same individual, it is certainly unclear whether any particular adverse reaction is due to the experimental therapy or to another cause. For example, suppose one of the volunteers became nauseous after all the patches had been applied to his/her back. It would be impossible to relate this event to any one of the compounds. On the other hand, if the reaction is localized, such as irritation or swelling at an injection site or site of application in the dermal study, then it can easily be concluded that this reaction is associated with that treatment. Thus, for the purposes of safety evaluation, studies such as these in which individuals receive multiple simultaneous treatments pose their own challenges to the study investigators. Investigators using such designs should take special care in their stated objectives relating to treatment safety.

The purpose of the above examples is to illustrate that the experimental unit in any particular study must be carefully defined. It is not always at the animal or person level. Only by carefully examining how treatments are actually allocated to units can one make this determination. In studies with multiple levels of sampling, this issue plays a critical role in the estimation of sample size.

3.5 Randomization

Randomization is a key feature of interventional studies. If experimental units differ only with regard to the treatment conditions applied to them and if their experimental outcomes are substantially different, then this allows one to infer that the different treatment conditions caused the different outcomes.

In our context, we will define randomization as the process of allocating experimental units (patients, animals, or other) to treatment groups or conditions according to a well-defined stochastic law. Here by the term *stochastic* we mean that it involves some element of chance, such as picking numbers out of a hat, or preferably, using a computer program to generate appropriate randomized treatment group designations. Randomization should and will often apply to other processes during a study. For example, one would often want to randomize the order in which biological samples collected during the study are assayed.

Suppose, for example, that we are randomizing a total of 100 patients into a clinical study with four treatments (A, B, C, D) according to a completely randomized design (see Section 5.1). Then the randomization process should generate an assignment for each of the 100 patients so that exactly 25 will be assigned to each treatment. All possible orderings satisfying this condition will be equally likely.

Randomization applies to other designs as well. For example, in the randomized block design comparing the five dermatological compounds, each subject's test areas (A through E) are assigned randomly to each of the five compounds.

Instead of using a mechanical or physical process for performing the randomization (such as coin flips or even selecting numbers from a published table of random digits), it is usually preferable to devise randomization schedules by computer generation. First, use of computer programs allows one to generate large randomizations without difficulty. Second, this method will be less error prone than would a mechanical approach. Third, and perhaps most important, the process will be easily documented because a printout of the randomization can be kept as part of the study records.

A main benefit of randomization is that it automatically balances groups (in the long run) on unimportant or unobservable variables. If we know of one or several critical variables, we should take these into account in designing our study. However, we will often be unaware of other variables or be unable to control them. Because randomization makes each possible allocation of treatments to experimental units equally likely, we would expect that average values in each of the treatment groups for any particular variable would be approximately equal. This is not to say that there will never be instances of differences. In fact, we expect such differences to occur occasionally. However, such differences will occur relatively infrequently and if we really believe that they may be important, we can take them into account (after the fact) by incorporating them into a statistical analysis.*

As another example of randomization, suppose that one is considering the scheduling of assays for an experimental study. In the study, four treatments (A, B, C, D) are administered resulting in 16 samples per treatment for a total of 64 samples to be analyzed. Suppose that it is possible to analyze 16 samples per day, so that 4 days are required to perform the laboratory analyses. The worst design would be to schedule one treatment (A, B, C, or D) per day. Then assays for each treatment would be completely confounded with day of assay. A much better design would be to randomize treatments in blocks of four, with four blocks per day, as in Figure 3.1.

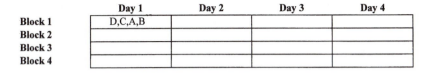

	Day 1	Day 2	Day 3	Day 4
Block 1	D,C,A,B			
Block 2				
Block 3				
Block 4				

Figure 3.1 Suggested design for laboratory assays of samples. Note: each cell should contain a random ordering of the four treatments (A, B, C, D).

Often, things do not work out as easily as they do in this theoretical example. Suppose, for instance, that only 15 samples can be run per day instead of 16. Here, I might suggest utilizing 3 runs out of the 15 as controls or a reference standard and the other 12 for the actual treatment assays. One would then randomize the four treatments plus reference (A, B, C, D, reference) into blocks of size five with three blocks per day. Of course, this will mean that the study assays will take 5 days to complete instead of 4, but this would be the case anyway, as only 60 could be done in the first four days. These types of tradeoffs should usually be considered in the planning of studies.

In multicenter clinical studies, it is generally advisable to randomize each site (center) separately. This ensures that treatment allocations will be balanced within each center as well as across centers. In practice, I usually like to randomize in blocks, where the blocking factor is two times the number of treatments. For example, if there are three treatments in the study, blocks of six would be utilized and each block would contain two subjects getting each treatment. Thus, in each study

---

* Analysis of covariance is one technique which is often utilized to compare different treatment regimens after adjustment has been made for possible differences at baseline. The variable which differs is called the *covariate* in the analysis.

center, there would be complete balance among treatments after each block of six patients is evaluated. If recruitment is substantially different from center to center, treatment allocations will still be approximately balanced with this approach.

In summary, randomization is an objective, scientifically accepted method for the allocation of experimental units to treatment groups. Randomization is encouraged by regulatory agencies and is easily documented.

3.6   Uniformity Trials

Uniformity trials are useful in identifying and quantifying sources of bias and variation without any interference from differing treatment conditions. They help to determine which factors need to be studied further to operationally remove or reduce bias and variability. Factors which contribute most to overall variability will also need the most replication. A uniformity trial consists of performing an actual study, but without applying differential treatments. For example, if a typical study involves administering test compounds or vehicles to laboratory animals, then the uniformity trial would involve administration of only the vehicle to all animals in the study in the same manner as usual. All other additional factors would be as they normally were and studied to assess whether differences in study outcomes among different levels of the factors were large enough to warrant further investigation.

In a uniformity study in the laboratory technician example discussed in Section 3.2, technicians could be one of the factors under study. Do the two technicians arrive at substantially different results even when all animals are administered vehicle?

Example 3.4 (Continuation of Example 1.1)

As an example of how a uniformity trial was useful in assessing bias and variability in an actual laboratory experiment, I will relate an interesting experience from several years ago. We were contacted by an investigator who had noticed what appeared to be considerable variability in many of his studies. He was performing experiments on mouse motor activity, which was a screening program to help assess the central nervous system (CNS) effects of new chemical entities by measuring their effect on the activity of mice in time intervals of fixed duration. It is well known that many drugs possess sedative or excitatory properties. These are often undesirable if the drug under study has clearly different purposes (e.g., lowering blood pressure).

This particular test system utilized a movable platform under each mouse's cage. As the mouse moved from one end of the cage to another, the platform rocked slightly and this rocking motion was detected and movement counts were accumulated over time. The counts in fixed time intervals were considered the outcomes of the study. In practice today, although such activity experiments are still conducted, counts are generated by the interruption of infrared beams of light arranged in a figure eight pattern in each cage.

The experimental setup consisted of a rack of 24 cages with four rows and six cages (columns) per row as depicted in Figure 3.2. In order to examine the investigator's claim of "excessive" variability, we decided to conduct a uniformity trial by performing a study free of drug treatments, i.e., only a vehicle preparation was given to each mouse in the study. Several other factors of interest may have been studied in addition, but the remarkable findings from this study were related only to cage locations.

Row 1 Column 1					Row 1 Column 6
Row 4 Column 1					Row 4 Column 6

Figure 3.2 Cage layout for mouse motor activity studies.

In particular, we found that mice in the first row (the highest level of the rack) had considerably lower activity counts than did mice in all other rows. This finding was obtained through statistical analysis of the uniformity trial data, but because the difference was so striking, it could have been seen easily simply by examination of the raw data and limited tabulations of them. When we alerted the investigator about this phenomenon, he pondered over it for a while, but then suggested that one possible explanation might be the difference in amount of light available in cages on the top row relative to all other rows. As rodents are nocturnal, most of their activity occurs when it is dark. Perhaps there was a substantial difference in lighting in the rack.

We then took light meter readings at all cage locations. Indeed what the investigator suspected was true! A new lighting system was installed in the animal room, one in which light intensity was almost identical throughout the rack. We performed another uniformity study after the installation. We found no consistent differences among rows or columns. Further, the variability among cages was also reduced substantially.

This example points out several interesting aspects of uniformity studies. First, although the investigator initially believed that his problem was one of excessive variability, it was actually one of potential bias, especially if all animals assigned to one of the treatment conditions were in the first row of the rack. Second, while it is not always possible to discover and remove the real source of bias and variability, this is clearly the most desirable approach to resolution. If we had not been able to adjust the lighting pattern in the room, we could have designed the study in a different way to overcome the problem. For example, we could have considered each row of the rack as a block and randomized each of six treatments to one of the six cage positions in each row. Block 1 would be very different from the other blocks. Moreover, we would be restricted to running exactly six compounds at a time (five test drugs plus vehicle) with exactly four experimental units per compound. Alternatively, we could have restricted the experiment to the bottom three rows only. This would clearly have been very inefficient as one fourth of our cages would have been empty at all times.

3.7   Blocking

If one or more factors other than the treatment conditions have been identified as potentially influencing the experimental outcomes, it may be appropriate to block or stratify on these factors. Units within a block are expected to act more alike than units in different blocks. For example, most toxicology studies randomize males and females separately and many preclinical animal studies block on the basis of initial weight of the animal.

If we are blocking on more than one factor, then the total number of blocks will equal the product of the number of levels of all factors. If we are blocking on both sex and initial weight range, for instance, and the number of weight strata is three, then there will be a total of six blocks in the study (two sexes × three weight intervals). Here, the weight intervals defining the blocks need not be the same for males and females. Experimental units should be randomized separately within each block or stratum.

In clinical trials, blocks are often defined by disease severity, prognosis, and/or demographic characteristics. In animal studies, blocking is often performed by sex, weight range, and/or other attributes at the start of the study. In engineering studies to assess the effects of differing conditions on a process, blocks may be defined by operators, machines, etc. Several real-life examples will help to illustrate.

Example 3.5

In a clinical study of the effects of a drug on the treatment of migraine, patients were to be randomized to either drug or placebo. Patients were first stratified on the basis of average number of migraines in the previous 3-month period and randomized separately within each stratum.

Example 3.6

In a veterinary clinical study of a temporary oxygen carrying blood replacement, dogs were randomized to active therapy (drug) or to a monitoring group (control). Patients were first stratified according to cause of anemia (blood loss, hemolysis, ineffective erythropoiesis) and randomized separately within each of the three strata.

Because units in the same block or stratum are expected to behave more alike than units in different strata, they will likely respond similarly to treatment. For example, the data in Table 3.3 illustrate the effect of stratification on the basis of a single factor. Within each cell, the range is exactly two units (i.e., the difference between the larger and smaller value). Suppose that we had not stratified on the basis of sex, but had observed the exact same eight outcomes as in Table 3.3. The difference within each treatment group (column of the table) and hence the "unexplained variation" would be much larger. By incorporating the factor of sex into the design of the study (i.e., by performing a randomized block design instead of a completely randomized design), we have substantially reduced the unexplained variation in the study and hence, improved its precision.

Table 3.3 Sample data with stratification by sex

	Treatment A	Treatment B
Males	156, 158	166, 168
Females	124, 126	132, 134

Even though we would have had a much poorer experiment without blocking due to the increased variability, an even worse outcome could have occurred. Suppose that our randomization had resulted in all or almost all males being in one treatment group and all or most females in the other. Then part of the "apparent" difference due to treatments would in fact be due to sex differences in the groups. Again, we would have complete or partial confounding. Therefore, when we can identify important factors which may strongly affect experimental outcomes, it is imperative to balance treatment assignments within these factors through the use of blocking.

3.8   Blinding

Blinding is a useful device to minimize study bias, especially when some endpoints are subjectively evaluated. In Section 3.2 we discussed the idea of blinding evaluators (raters) as to the treatment condition of the experimental subjects. In addition, many studies utilize blinding for the subjects themselves (i.e., if the subjects are people) and blinding for investigators. In clinical studies, the former condition is called single-blinding and the latter is called double-blinding, when both subjects and investigators are unaware as to the treatment designations of the subjects.

For safety reasons, it is important to be able to quickly "break the blind" if a serious or life threatening condition arises for a particular subject. Most protocols identify a study monitor (with telephone number) who possesses such information and can readily respond in case of emergency.

It is well known that many disease states are susceptible to placebo effects. Just by administering a supposedly active treatment to a subject, his/her awareness of the disease may be altered and some apparent benefits of therapy may be perceived. Similarly, subjects sometimes experience "side effects" due only to their anticipation, with psychosomatic causes. What would happen if subjects knew they were on active treatment or placebo? Those on active treatment might be more inclined to perceive a benefit of therapy and/or side effects from the treatment. Those subjects on placebo would likely be less inclined to perceive both. Here we would have a potential bias introduced which could be eliminated by single-blinding.

In order to make the blinding work, all treatment groups must be handled as identically as possible. For example, if active treatment is administered to patients in a clinical study via tablets or capsules, then all placebo-treated patients should also be administered tablets or capsules which match those for active treatment by size, color, packaging, etc. Identifying information should be coded so that the blind can be broken in emergency situations, but should not otherwise be used to identify treatment group.

Similarly, it is often advantageous to blind investigators as to the treatment status of experimental subjects. This is especially true when some outcome variables are subjectively determined by the investigators (e.g., physician scores of patient condition), but blinding should almost always be utilized even when objective endpoints are being used. Even the most objective scientist might unknowingly slightly alter his/her ratings or measurements based on knowledge of treatment group and its supposed effect. Besides, why permit this possibility and allow for potential criticism of an otherwise excellent study?

Blinding also prevents selection bias which is not a trivial problem in clinical trials. By not knowing the order of treatments, a physician cannot selectively enroll patients to favor one treatment or another.

Clearly, it is not always possible or practical to blind subjects and/or investigators. For example, in a clinical study to compare the effectiveness of removal of gallstones by surgical or chemical means, it would be foolish to consider subjecting patients in the "drug group" to unnecessary sham surgery. Despite such limitations, blinding remains an essential mechanism for minimizing bias in experimental studies.

chapter four

Sample Size Estimation

The process of estimating appropriate sample sizes for biomedical studies is really a straightforward one, where the estimation depends on the statistical context, the assumptions being made, and the specifications for the study. Both the context and the specifications will depend on the study objectives and design. By statistical context, I refer to whether the primary purpose of the study (as it relates to sample size estimation) is for point estimation, interval estimation, or hypothesis testing. These concepts are specifically defined in Section 4.1 along with their relevant characteristics. As each of these three situations is somewhat different, and the terms needed to estimate sample sizes vary from one context to another, they are described in further detail in Sections 4.2 through 4.4. Typically, one needs to make certain assumptions about the data that will be collected in a biomedical study. For example, in order to estimate sample sizes in most contexts, one needs some prior information about the precision or variability of the data. Such information may be available from published literature where similar studies have been conducted, or from small preliminary studies (i.e., pilot studies) which have as one of their major objectives the estimation of variability for the primary or pivotal study (see Section 4.5). As discussed earlier in Section 3.3, many studies contain situations in which there are multiple levels of sampling, with those performed at lower levels called subsampling (or even sub-subsampling). Here, we often want to perform sample size estimation at each possible level in order to maximize the efficiency of our design (see Section 4.6). Finally, because sample size estimates are subject to assumptions and particular specifications, we may wish to perform a sensitivity analysis to see how sample sizes change with differing assumptions and/or specifications (Section 4.7).

4.1   Statistical Context

In considering the estimation of sample sizes for a biomedical study, we must distinguish which of several statistical contexts apply. In this chapter, we will specifically consider three possible contexts: point estimation, interval estimation, and hypothesis testing. These are perhaps the most frequently encountered in practice and represent the vast majority of applications.

Each of these contexts is of course related to the objectives of the study. One must decide how the primary objectives of the study relate to these and often, which of several possible response variables will be selected to determine the sample size calculation. In practice, if there are several candidate variables and each is relatively equally important, then the response variable generating the largest sample size will be selected as this will ensure that the criteria for all others are satisfied as well.

In point estimation, the final result of interest will be a single quantity called a sample statistic which is intended to estimate a true population quantity called a population parameter. For example, a clinical study may be primarily concerned with estimating the 5-year survival rate of patients with a particular type of cancer given a new experimental treatment. The idea is to precisely estimate the 5-year survival rate. The final result in such a context is often expressed as the mean (or sample estimate) plus or minus (±) the standard error of estimation. This gives some idea as to how precisely the estimate from the sample (in this case the set of patients studied in the clinical trial) estimates the true population (i.e., all patients who might ultimately receive this therapy under similar conditions) parameter (the overall 5-year survival rate among all such patients). Probably the two cases of greatest practical interest are when the data are dichotomous (i.e., have only two possible outcomes such as success or failure) and when the data are continuous (as distinguished from data that are discrete). Standard errors are larger with more inherent variability and become smaller as the sample size increases. One chooses a sample size in order to make the standard error of estimation sufficiently small (i.e., make the estimate itself sufficiently precise).

In interval estimation, the end result is intended to be an interval such as [A,B], with lower limit A and upper limit B, which contains the true population quantity with predetermined probability. This is called a confidence interval. In fact, the interval either contains the true population quantity or it does not contain it. The idea is that this interval depends on the sample data and hence is random (because the sample data are subject to variation). By this I mean that the outcome depends on the observed sample and therefore is subject to variation and uncertainty. We cannot say for certain beforehand, however, whether such a random interval will or will not contain the population quantity considered as fixed but unknown, only that it will do so with the prespecified probability over the long run (i.e., over many such experiments). The appropriate sample size will again depend on the inherent variability in the data, and also on the probability of coverage. This probability is called the confidence coefficient. Typical confidence coefficients are high, such as 90, 95, or 99%.

Perhaps the most common statistical context in biomedical studies is hypothesis testing. One specifies a particular null hypothesis which will either be accepted or rejected at the conclusion of the study. In situations where different treatment conditions are applied either within or between experimental units, the null hypothesis will often be that the response variable does not depend on the treatment condition. For example, one may state as a null hypothesis that the population means of a particular parameter are equal under two or more different treatment conditions. Similarly, another null hypothesis of interest may be that response probabilities (i.e., the probability of a positive response such as a successful patient outcome) is independent of treatment condition. Another null hypothesis might be that survival experiences in several different populations are independent of the specific population (treatment condition). Because the specific treatment condition does not provide us with real information about likely outcomes, these situations are referred to as "the null hypotheses." In order to estimate sample sizes in hypothesis testing, however, one needs to specify not only the null hypothesis, but also an alternative hypothesis which is presumed to hold if the null hypothesis does not.

4.2   *Sample Sizes for Point Estimation*

In order to achieve sufficient precision of estimation, one must use some measure of variability as the criterion. Most often, the final estimate will be accompanied by its standard error which is one estimate of its uncertainty. Therefore, in point estimation, the most common criterion for sample size determination is the standard error.

The basic idea behind sample size estimation when the criterion is the standard error is that the standard error of the estimate decreases as the sample size increases. The reason for this is that larger samples provide us with more information about the population. In fact, the standard error of an estimate is inversely related to the square root of the sample size (see Appendix B.1 for specific formulas). In addition, sample sizes will of necessity be larger with larger inherent variability as well. In order to determine a sample size, one needs to estimate this variability (based on data from the literature or from pilot studies), and then calculate the sample size to achieve the desired precision. Usually this is expressed as a percent of the estimate itself, e.g., the criterion for the standard error might be ± 5% or ± 10%.

Example 4.1

As an example, one aspect of a large study in rats had as its goal the characterization of the pharmacokinetic profile of a drug. Because only a single blood sample could be obtained from each rat in the study, estimation of quantities such as the area under the plasma concentration–time curve (AUC) involved use of the mean concentration for all rats sampled in a group (drug level). Using data from a previous study to estimate the expected variability in the current study, we calculated that eight rats per time point (seven time points) would suffice so that the anticipated standard error of the estimated AUC would be no more than 10% (relative to the actual AUC).

4.3   Sample Sizes for Interval Estimation

While a point estimate is useful as a single number describing a population quantity, it is often desirable to be able to make statements about such parameters at a specific level of probability (confidence). When the statements about these parameters are of the form of an interval which contains the true parameter with prespecified probability, the intervals are called confidence intervals (as defined above) and the statements are called confidence statements.

What would be desirable properties of confidence intervals? First, we would like to contain the population parameter with a high probability or degree of confidence, typically 90% or more. Second, we would like such intervals to be as narrow as possible. To illustrate this latter idea, suppose that the population parameter we are estimating is the 5-year survival rate following a new cancer treatment. What would we learn if we could say that survival was between 10 and 90% with 95% confidence? We would not learn much, as these limits are too wide to be of any practical use.

Unfortunately, the two criteria for sample size estimation based on confidence intervals work at cross purposes. In order to have a very high confidence level, the confidence interval will of necessity be wider in order to have a higher probability of containing the true parameter. Similarly, if a confidence interval is narrow, it will have a lower probability of containing the true population parameter. In determining sample sizes, one often considers the tradeoff between confidence level and the expected width of the confidence interval and tries to make the best compromise between the two. Specific formulas for sample size estimation with confidence intervals as their basis are provided in Appendix B.2.

Example 4.2

As an example, one possible design considered for the clinical evaluation of a new biological product was a comparison with another similar therapy. We would conclude that the two products were equally effective if the difference in success rates could be established to be small. Another way of saying this would be that we could determine with a given degree of confidence that the true difference in success rates would be within a few percent. Thus, sample size calculations were performed using a confidence interval as the basis. In particular, we considered a confidence level of 95% and percent differences of 5 and 10% between the two products. In addition, the sample size calculation (see Appendix B) required that the 95% confidence interval would be sufficiently narrow (i.e., up to 5% or up to 10%) with 80% probability (see Makuch and Johnson, 1990). The resulting sample size estimates are displayed in Table 4.1. For example, in order to have an 80% chance that the 95% upper confidence limit on the difference in success rates be less than 10% (assuming a common success rate of 85%) would require 157 patients per group, for a total of 314 patients.

Table 4.1 Sample sizes per group for active controlled study (based on 95% confidence coefficient; 80% probability that limit is less than selected bound)

Common success rate (both therapies)	Selected bound on difference	
	5%	10%
80%	791	198
85%	630	**157**
90%	445	111

As it turned out, we decided not to pursue an active controlled trial for this product, but rather to pursue a completely different design. By performing the above sample size calculations, however, we were able to consider this alternative and weigh its merits relative to other designs.

4.4   Sample Sizes for Hypothesis Testing

In order to describe sample size estimation in the context of hypothesis testing, we must first define some terms which are used in the hypothesis testing context. As indicated above, we must first identify a null hypothesis of interest and, for the purposes of sample size estimation, an alternative hypothesis of interest. At the end of the study, we will either accept or reject the null hypothesis. Whether this is correct or not depends on whether the null hypothesis is actually true or whether it is false.

As indicated in Table 4.2, there are four possible outcomes at the end of the experiment. When the null hypothesis is true and one accepts the null hypothesis (i.e., fails to reject it), then one has made the correct decision. Similarly, if the null hypothesis is not true and hence, some alternative hypothesis holds and one rejects the null, then one has also made the correct decision. However, there are two erroneous possibilities. If the null hypothesis is true, and one incorrectly rejects it, then one has made an error of a *false positive*. Alternatively, if some alternative actually holds yet one fails to reject the null hypothesis, then one has made an error of a *false negative*.*

Table 4.2   The decision process in hypothesis testing

State of Nature (Truth)

Decision made	Null hypothesis true	Alternative hypothesis true
Accept null hypothesis	Correct decision	False negative
Reject null hypothesis	False positive	Correct decision

Clearly, it is desirable to limit the rate at which such errors are made. In fact, the basis of sample size calculations in hypothesis testing is to specify an allowable rate of false positives and an allowable rate of false negatives (for a particular alternative hypothesis) and to estimate a sample size large enough so that these low error rates can be achieved. The allowable rate of false positives is called the *level of significance* (sometimes referred to as the *alpha level*). In describing the rate of false negatives, this is often described in terms of the probability of rejecting the null hypothesis when the postulated alternative holds (i.e., and a correct decision is made). This is called the *power* of the statistical test of significance.** The alpha level or level of significance is often set at values such as .01, .05, or .10. Power levels are usually at or above 80%, e.g., 80, 90, 95%, etc.

As with the two criteria for sample sizes with confidence interval estimation, the two criteria for sample size estimation in the hypothesis testing context (false positive rate, power) work at cross purposes. One can make the false positive rate low by making it very difficult to reject the null hypothesis. In an informal way, we can say that this means

---

* The terms false positive and false negative are often referred to as "Type I errors" and "Type II errors," respectively. However, I find these latter terms much less descriptive and potentially confusing.

** A plot of the power of the test vs. the magnitude of the true treatment effect is called an *operating characteristic curve*.

that we will require quite a bit of evidence that the null hypothesis is false in order to reject it. However, if we do this, then it will be difficult to reject and hence, the study will have low power. Conversely, if we achieve high power by making it easy to reject the null hypothesis, we will inflate the false positive rate. In estimating sample sizes in the context of hypothesis testing, one usually strikes a balance between the false positive rate, the power of the study, and the sample size. While we would love to have a study with a very low false positive rate and very high power, this will usually require a very large sample size which may not be practical. In designing studies, we can assess the costs of various tradeoffs in these quantities and select the combination that is most suitable (see Section 4.7 regarding sensitivity analyses for sample size estimation).

How does hypothesis testing work? In addition to the terms described above and the rate of allowable false positives set prior to the study (the alpha level), the idea of hypothesis testing is accomplished through what is called a *test statistic*. The concept is that extreme values of the test statistic are unlikely when the null hypothesis holds. Therefore, when we observe such extreme values, this would generally indicate that the null hypothesis does not hold and we should reject it. However, if we observe values of the test statistic which are not extreme, then there would not be enough evidence to lead to rejection of the null hypothesis.

The test statistic is a quantity calculated from the experimental data and for which values of the quantity are somewhat indicative of whether or not the null hypothesis holds. A few examples will help to illustrate this idea. Suppose, for example, that one is testing two alternative therapies in a clinical situation, and one can characterize the results of therapy as either a success or failure. A suitable null hypothesis in this example would be that the success rates of the two therapies were equal and a test statistic might be based on the observed difference in success rates in the study. A second similar example might involve the population means for two different experimental conditions. Here an obvious test statistic might utilize the observed difference in sample means from the study data.

In each of these two examples, we have not indicated how the alternative hypothesis was specified, however. In fact, this is an important issue in determining which values of the test statistic are extreme and the consequences of this on sample sizes. In considering the difference in success rates between the two therapies, will the alternative hypothesis be that Therapy 2 is superior to Therapy 1 (i.e., has a higher success rate) or that there is a difference in either direction (i.e., one therapy is superior to the other, but either one may be the superior one)? The former situation is referred to as a *one-sided hypothesis* and the corresponding statistical test is called a *one-sided test*. The latter situation is referred to as a *two-sided hypothesis* and the test is then a *two-sided test*. One must specify in advance whether the testing procedure is one-sided or two-sided, and this will affect the sample size estimation (see Appendix B.3 for specific formulas).

What do we mean by extreme values of a test statistic? We mean values of the test statistic that are unlikely or improbable under the null hypothesis. In order to quantify this concept, we must be able to determine the probability of a test statistic as extreme or more extreme than the actual one we observe in the study. For example, suppose we observe a success rate for one therapy of 85% and a success rate for the other therapy of 70%. Is this difference large enough so that it is improbable (has a low probability) when the two therapies have a common rate of successes? Part of the answer will depend on whether we expected the difference to be in the direction as indicated by the data beforehand (i.e., we were in a one-sided testing situation) or whether a difference in either direction would have been of interest (a two-sided alternative).

The calculated probability of a test statistic as extreme or more extreme than the observed one (under the null hypothesis) is called the "*p*-value." When this probability or *p*-value is less than the alpha level, one rejects the null hypothesis. When the *p*-value exceeds the alpha level, then the null hypothesis is accepted.

One other set of terms is often used in describing statistical hypothesis testing. Because a null hypothesis is rejected for extreme values of a test statistic, these values are unlikely under the null hypothesis and hence, are values in the "tails" or extreme parts of the underlying probability distribution. Accordingly, one-sided tests are also called *one-tailed tests* and two-sided tests are also called *two-tailed tests*. One-tailed tests will reject when the value of the test statistic lies in the upper(/lower) tail of the probability distribution. Two-tailed tests reject when the value of the test statistic is either very high (in the upper tail) or very low (in the lower tail). The cutoffs or "critical values" are determined in such a way that the overall false positive rate is set at alpha (usually .10, .05, or .01). In most cases, the error rate is allocated equally between the upper and lower tail for two-sided tests.

In general, sample sizes for hypothesis testing will depend on four factors: the inherent variability in the study parameter of interest, the size of the difference to be detected, the allowable rate of false positives (level of significance), and the power one wishes to achieve. As always, sample sizes will be larger when the variability is larger. In addition, sample sizes will also be larger when the difference to be detected is small. Conversely, if the difference to be detected is large, it will be easier to do so, and hence, the required sample sizes will be less. The difference to be detected is reflective of the state of nature under the alternative hypothesis. This difference should be chosen in order to be most clinically or biologically relevant. One should always ask, "How large a difference would it be important to be able to detect?" For example, if one is studying potentially increased 5-year survival rates as a function of a new cancer therapy, how great an increase would be medically important relative to the current survival rate under standard therapy? The effects of level of significance (alpha level) and power on sample sizes have been discussed above.

Example 4.2 (continued)

As an example of sample size estimation in the context of hypothesis testing, again consider the clinical study for the biological product mentioned above. Another possible design was a comparison against a placebo control. We believed that the efficacy rate would be over-whelmingly in favor of active therapy. Sample sizes, therefore, were generated in order to be able to assess (test for) a slightly higher rate of side effects with active therapy.

The specific sample size estimates generated in this way are displayed in Table 4.3 and were obtained from Appendix Table A.3 of Fleiss (1981). For example, in order to be able to detect a difference in adverse drug reaction rates between 5% in the placebo group and 15% in the treated group, this would require 160 patients per group (320 total) to achieve 80% power using a two-sided test at the 5% level.

Table 4.3 Sample sizes per group required for placebo controlled study (80% power; two-sided test at 5% level)

Rate of ADRs in placebo group	Rate of ADRs in treated group			
	10%	15%	20%	25%
5%	474	**160**	88	58
10%	—	725	219	113
15%	—	—	945	270

4.5 Pilot Studies

Pilot studies are especially useful in cases in which we are undertaking very time-consuming, important, or expensive studies. There are several purposes for pilot studies. A primary purpose is to work out and refine experimental or technical procedures that will be used during the larger subsequent study. A second purpose is often to train study personnel by allowing them to get experience with such techniques particularly when the techniques or therapies are substantially different from those with which they are familiar.

Pilot studies are often useful in getting a preliminary assessment of the potential efficacy of a new product or therapy. For example, are the benefits from a new drug or experimental therapy likely to be large enough to warrant further study? While we may not be able to answer such questions definitively, we may be able to get a preliminary idea of efficacy and this may be useful in helping us to make a strategic decision.

Another important use of pilot studies is to provide information about anticipated variability in the future comprehensive study. Notice that all the sample size formulas provided in Appendix B require either estimates of variability (continuous cases) or estimates of the underlying rate(s) or frequencies (for dichotomous cases). Sometimes these estimates are available from literature references describing identical or similar studies previously conducted, but more often, such estimates are not available and must be generated based on pilot data collected with this objective in mind.

While it is often difficult to reconcile the varied objectives in a pilot study, it seems critical that if one of the uses is to provide estimates of variability for sample size estimation, then at least a portion of the study should be conducted in a manner that simulates the experimental protocol as closely as possible. By this I mean that although one may be refining experimental techniques throughout the conduct of pilot studies, the portion of these studies which are utilized for sample size estimation should reflect the proposed experimental protocol techniques as closely as is practical.

Finally, it should be remembered that data collected in pilot studies are also subject to variability so that estimates derived from these data will also be subject to uncertainty. This is particularly critical as pilot studies will almost always be substantially smaller than the proposed studies themselves. Therefore, one should make allowance for this fact when generating sample size estimates, perhaps by conducting a sensitivity analysis (Section 4.7) with a range of values consistent with those observed in the pilot study.

Example 4.3

As an illustration of this idea, consider a situation in which the primary outcome variable in a proposed study is dichotomous, such as one in which experimental units are ultimately classified as either successes or failures. As depicted in Figure 4.1, it is well known that the most variability for a single experimental unit occurs when the true success rate is at 50%. Accordingly, the highest variability of the sample success rate (total number of successes divided by number in group) in a group of experimental units will also occur when the underlying rate is 50%. The most conservative estimate of sample size will be based on a 50% success rate and this approach has been recommended elsewhere (Kupper and Hafner, 1989). However, I believe that this approach may lead to very large sample sizes, especially when it is expected that the true success rate is very low or very high (close to zero or close to one; see Figure 4.1). It may be difficult, however, to accurately estimate a true rate based on a small pilot study. For example, if a pilot study is conducted with ten experimental units and two successes are observed, even this outcome is consistent with a success rate of 50% or more.

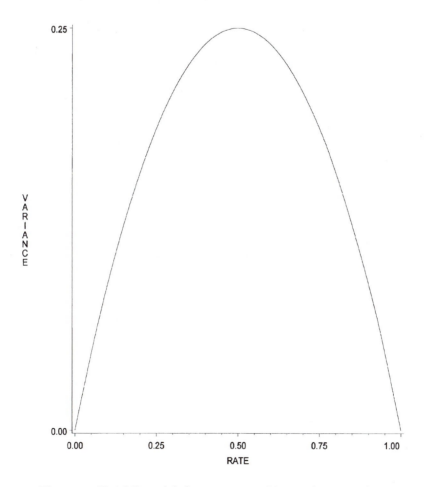

Figure 4.1 Variability of dichotomous variable as a function of rate.

Therefore, it seems that the best strategy is to consider the kind and utility of information being obtained in a pilot study. At a minimum, estimates of subject to subject variability in the continuous case should probably not be based on less than six or so experimental units, although even then, one needs to consider the uncertainty in such estimation. In the dichotomous case, larger sample sizes are necessary to provide a satisfactory estimate of an underlying rate variable. A good compromise in such situations is to perform a sensitivity analysis and then to select a relatively conservative sample size taking into consideration issues of cost, study duration, and the present state of knowledge.

By conducting pilot studies in this way and generating sample sizes accordingly, one will hope to avoid the frustrating outcome where one conducts a large, expensive, and/or critical study, but at the end finds out that the results are "not quite definitive" and that a slightly larger or more efficient study was needed. Finally, we remark that, in some circumstances, the issue of a single fixed sample size can be avoided by conducting a study sequentially (see Chapter 6), although this is not always a practical alternative to a fixed size design.

4.6 Subsampling Issues

In some situations, there are multiple levels of sampling. In animal studies, for example, in which animals are multiply housed or caged and all animals in the same cage are administered the same treatment, the cage is the primary experimental unit, and the animal may be regarded as a secondary or subsampling unit. Similarly, in many *in vitro* studies, treatments are administered on a plate or test tube basis, whereas data are often recorded at the colony or cell level. In experiments such as these and others, the issue of sample size estimation is further complicated by the various levels of data being collected.

How should one view sample size estimation when there are multiple levels of sampling? In contrast to the case of experimental units at a single level, there can be a variety of costs associated with each different sampling level. For example, in an *in vitro* study with replicate plates per treatment condition and colony counting or assessment as a factor, there is the setup time for individual plates and the time and effort to assess individual colonies. In deciding how many replicate plates and how many colonies per plate need to be assessed, one must determine the relative amount of information (i.e., precision of estimation) provided by each level. Often there is a tradeoff between the factors at various levels, so that the same amount of information may be provided by different combinations of sample sizes at the various levels. This concept is illustrated for the study of hepatic cells in rats discussed in Section 3.3.

Example 4.4

An illustration of estimation for samples and subsamples in the context of hypothesis testing was provided by Whorton et al. (1979). Whorton et al. studied cytogenetic aberration rates in a population of potential workers at a division of a large U.S. chemical company. The goal of their studies was to assess a baseline rate of abnormal cells and to estimate required sample sizes to test for changes in such a population compared to a concurrent control group.

In the baseline study, Whorton et al. found that the proportion of abnormal cells was about 2% for the population studied and also that the between-subject variance of abnormal cells was significantly greater in this population than the within-subject variance. Accordingly, they considered sample size estimation with two levels of sampling: subjects and cells within subjects. In particular, it was determined that, in order to detect a change from baseline of 2% (i.e., from 2 to 4%) with a power of 80% using a one-sided 5% test, 14 subjects were required per group with 200 cells assessed per subject. The total number of cells studied therefore would be 14 × 200 = 2800 per group. Alternatively, if 19 subjects were included in each of the control and test group, only 100 cells would be required per subject for a total of 1900 cells. Thus, if most of the experimental effort would be expended in assessing cells, the total number of cells and hence, the total effort, could be reduced substantially by sampling more subjects and fewer cells per subject. Examples such as this one illustrate the kind of tradeoffs that exist with multiple levels of sampling.

4.7   *Sensitivity Analyses*

Just as an interval estimate (Section 4.3) about a population parameter is often superior to a point estimate providing only a single value representation, it is often desirable to provide more than one estimate of sample size. When we vary some of the underlying assumptions (e.g., anticipated precision in the study) and/or specifications (e.g., alternative hypothesis, level, power in hypothesis testing) and conduct sample size estimation under several such configurations, we are conducting a sensitivity analysis. In general, sensitivity analyses indicate how results change based on modifications to underlying assumptions or specifications.

It is useful to conduct sensitivity analyses on sample sizes for several reasons. First, as discussed above, the information available about such items as precision and magnitude of likely effects are often preliminary at the time when sample size estimates are made. It is often helpful to see how sample sizes vary under slightly different conditions. Second, it allows one to perform a "what if" assessment to see how changes in various specifications affect the desired sample size. By doing this, it allows study planners to make the best informed decisions. Finally, by performing sample size calculations under a variety of conditions and even possible design alternatives, it helps one to make the best choice among competing experimental designs.

The kinds of sensitivity analyses that are often conducted are illustrated in Tables 4.1 and 4.3. In Table 4.1, a sample size calculation based on confidence intervals is exhibited. The two parameters that vary in the table are the common success rate for the two therapies (80, 85, or 90%) and the upper bound on the difference in success rates between the therapies (5 and 10%). Two other parameters are held fixed in this example: the confidence coefficient (set at 95%) and the probability that the bound will be achieved (set at 80%). In principle, one could vary these as well and assess the effect of a higher or lower level of confidence and a higher or lower probability. The main finding from this table is that it would be totally impractical to attempt to achieve a 5% bound on the difference in success rates and that even a bound of 10% would require relatively large sample sizes (compared to the target population size, likely recruiting rates, etc.).

In Table 4.3, we illustrate sample size calculations and a sensitivity analysis performed in the context of hypothesis testing. The two parameters that vary are the rate of adverse drug reactions (ADRs) in the placebo group (5, 10, and 15%) and the rate of ADRs in the treated group (10, 15, 20, and 25%). Two other parameters are held constant: the level of the two-sided test set at 5% and the required power of the test set at 80%. Again, these also could have been varied to assess the effects of changes in them on sample sizes. The main finding from this table is that, in the range studied, at least a 10% difference in adverse reaction rates would be necessary (e.g., 5 vs. 15%) to be detectable with the very maximum sample size anticipated for the study (about 200 patients per group). Even a larger difference would be necessary (say 15% or more) if the sample size was to be about 100 patients per group.

Producing only a single number as "the sample size estimate" is not only uninformative relative to a full sensitivity analysis, but it is also misleading. When an investigator is informed that "the statistician has determined that a sample size of X is required," this statement provides too much emphasis on this single number. In most situations, the amount and definitiveness of information available at the sample size estimation stage either from published literature or from pilot studies are usually not good enough simply to rely on them without further investigation. Moreover, by presenting only a single sample size number, one is really doing a disservice to the amount of information that can be provided at the planning stages of a study. By performing a full sensitivity analysis, one can generate enough and valuable data to assist study planners in making the best informed decision. Such decisions will not only consider the N for the study, but also allow one to consider tradeoffs between competing designs.

In practice, I often calculate sample sizes for a range of values for the difference to be detected, the assumed variability, and the power of the study. If it becomes evident that even with the largest practical sample size, the power of a study is likely to be low (e.g., 70% or less) then it becomes questionable as to whether the study should be undertaken at all. Consideration can be given to alternative studies (see Section 2.1) or to whether further preliminary research is needed to reduce experimental variability (see Section 3.3).

As with all data evaluations, one must be careful not to overwhelm his/her colleagues with so much information that they cannot digest most of it and tend to ignore all of it. It is therefore wise to restrict sensitivity analyses to the most important parameters and to consider the likely range of values for these. Such decisions about what and how much data to present are best made after consultations with subject matter experts.

Despite this mild caveat, it is clear that the use of sensitivity analyses to enhance sample size estimation is an important tool in the planning of biomedical experiments.

chapter five

Common Designs in Biological Experimentation

In almost all biomedical studies, there are alternative experimental designs which can be considered in the planning stages of the studies. The range of such designs will depend on the particular situation at hand and it would be impossible to specify completely which types of designs might be considered without firsthand knowledge of the specific details. However, we can identify several classes of experimental designs which are most commonly employed in biomedical research and present information about them.

Selecting the best design will depend on a number of factors. Some designs will be completely infeasible. For example, in a clinical study in which long-term survival is the primary endpoint, designs which incorporate crossover from one treatment condition to another at the end of a fixed period could usually be ruled out immediately. Patients who died during the initial period would not be available for subsequent periods and even patients who survived would be somewhat older and with perhaps a different disease state in later periods.

In selecting among alternative feasible designs, one should consider such issues as the number of experimental (and perhaps other) units predicted to achieve the desired result (as discussed in Chapter 4), duration and costs of the study, and other possible consequences of the design such as increased or decreased dropout rates relative to other designs. Designs which appear to be suitable in this regard should be proposed and their relative merits presented and discussed with the study investigators until a consensus choice is achieved.

Even within a class of designs, there will be many choices to be made. If the study is to be a controlled study, should the control be a negative or positive control, and if a negative control, what form should it take? How many doses are appropriate? What dose level or levels should be studied? What should the frequency of dosing be? What is the route of administration? And so on. Clearly the answers to many of these questions will depend on both the management and study objectives.

In the remainder of this chapter, we present and discuss several of the most common classes of experimental designs used in biomedical studies. As indicated below, each has its advantages and disadvantages relative to others and these factors dictate which classes can be considered for a given study.

5.1   The Completely Randomized Design

Perhaps the most common type of experimental design in biomedical research is the completely randomized design. In a completely randomized design, each experimental unit or subject is randomly assigned to exactly one treatment condition. In this case, the subject is said to be "nested" within a treatment group. If the treatment condition represents a single factor under study (e.g., drug at several levels, perhaps including a zero level), then the design is called a *one-way layout*. This term arises from the idea that the treatment conditions can be laid out in a table with just a single row as in Table 5.1. When the treatment conditions represent multiple factors, each at two or more levels, and with all such factors considered as "the treatment," then this design is called a *factorial design* as the joint effect of the several factors on outcome is experimentally assessed. These designs are not only efficient for determining the joint effect of several factors simultaneously, but are also efficient for each factor individually due to the so-called "hidden replication" in factorial designs.

Table 5.1   Treatment groups in a one-way layout
with K groups

Treatment 1	Treatment 2	...	Treatment K

For example, suppose that we are considering an experiment in which the effect of two factors, A and B, will be investigated. Suppose further that A will be studied at two levels: A1 and A2, and that B will be studied at three levels: B1, B2, and B3. There are $2 \times 3 = 6$ treatment combinations for this study as depicted in Table 5.2. Because there are two experimental factors under study (A and B), the design is called a two-factor factorial design. The hidden replication referred to above in this study obtains because each treatment is applied to several cells (e.g., A1 is applied in three cells corresponding to the three levels of

B). Balanced designs such as this, but with three factors (e.g., A, B, and C), would be called three-factor factorial designs, and so on.

Table 5.2 Treatment combinations in 2 × 3 factorial study

A1 and B1	A1 and B2	A1 and B3
A2 and B1	A2 and B2	A2 and B3

5.1.1 The One-Way Layout

The objective in a study designed as a one-way layout is to assess the effects of different levels of a factor on the response variable or variables. The factor can represent presence or absence of treatment (e.g., drug vs. placebo), several different related treatment conditions (e.g., Drug A, Drug B, Drug C each at a single level), or several levels of a single condition (e.g., no drug, low dose Drug A, mid dose Drug A, and high dose Drug A). Experimental units are randomly allocated to treatment groups in such a way that each group is equally likely. In addition, if the number of experimental units in each group is the same, then the design is called a balanced design.

Example 5.1

Table 5.3 depicts the allocation of laboratory animals in a dose-response study of a new experimental compound in a particular test system. Ten animals are randomized to each of four dose levels: control (zero dose), low dose, mid dose, and high dose, for a total of forty animals in the entire experiment. As can be clearly seen from the table, the study is an illustration of the one-way layout with four levels of the treatment.

Table 5.3 Dose-response study with four dose levels

Treatment 1	Treatment 2	Treatment 3	Treatment 4
Control (10 animals)	Low dose (10 animals)	Mid dose (10 animals)	High dose (10 animals)

5.1.2 Factorial Design

In situations in which experimenters are interested in assessing the joint effect of two or more factors, they can utilize factorial designs. The advantage of factorial designs over "one factor at a time" studies is that the response to simultaneous administration of the factors can be assessed. Factorial designs in which the total number of factors is relatively low (2, 3, or 4) are most common. As the number of factors and the number of levels per factor increase, so does the total number of possible treatment combinations. Because this total is multiplicative, i.e., the total number of levels of all factors is multiplied to yield the total number of treatment combinations, a full factorial study can easily become quite large. For example, if one is considering a three factor factorial study in which each factor is to be assessed at four levels, the total number of treatment combinations in the study would be $4 \times 4 \times 4 = 64$. If the study is a balanced design with each cell equally replicated with N experimental units per cell, then the total number of experimental units would be 64 N. Even with a relatively small N, this can be a large study!

Example 5.2

A common type of factorial design is one in which each factor is assessed at exactly two levels. If the number of factors is K, then such a design is called a 2^K *design*. For example, suppose that one is interested in assessing the joint and individual effects of two factors, A and B. The two levels of A will be A absent and A present. Similarly, the two levels of B will be B absent and B present. The resultant four treatment combinations will be as given in Table 5.4.

Table 5.4   Treatment combinations in
2 × 2 study

Control	A alone
B alone	A and B

A major objective for a study such as this would be to determine whether the response to the A + B treatment combination would be different or the same as the response that could be expected from the individual components alone. In particular, is the response to the joint administration substantially greater (better) than would be expected from the effects of the individual effect of A and the individual effect of B? Is the joint response substantially less (worse) than would be expected? In the former case, we say that the two factors exhibit synergy, whereas in the latter case, we say that the two factors exhibit interference.

In statistical terms, factors for which the joint effect is equal to the sum of the effects of the individual components are said to be "additive," as one can add up the effects of each component and thus get the joint effect. Factors which are not additive are said to exhibit "interaction" (synergy or interference). In particular, if the factors A and B are not additive, we say that there is an A × B interaction.

For the 2 × 2 study described above, an illustration of interaction is provided in Figure 5.1. The two solid lines in the figure indicate the "expected" outcome with additivity, i.e., when no A × B interaction is present. The joint effect of A + B (difference from control response) is the sum of the effect of A and the effect of B. It can be seen that the two solid lines are parallel and that the four points (CONTROL, A, B, A + B) form a parallelogram. Thus, the two differences (A + B – B) and (A – CONTROL) are the same. This indicates that the effect of A in the presence of B is the same as the effect of A in the absence of B. Thus the joint effect of A and B is the same as the sum of the effect of A (individually) and the effect of B (individually).

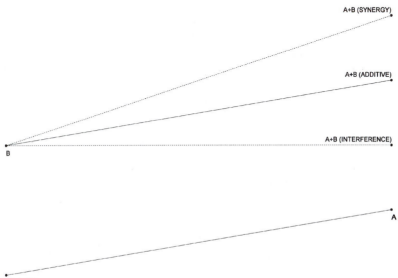

Figure 5.1   Illustration of interaction in a 2 × 2 factorial experiment.

The dashed line at the top of the figure illustrates an interaction (in this case synergy) between A and B. The effect of the combination is greater than would be expected from the individual components alone. The lower dashed line in the figure indicates another potential example of interaction. In this case, the effect of the combination (A + B) is lower than the sum of the individual effects of each component separately. In such a situation, we say that there is interference between the factors A and B. In either instance of interaction, we find that A and B are not additive, or that their joint effect does not equal the sum of the individual effects of each. A primary purpose of factorial studies is often to assess potential interactions between factors.*

---

* In contrast, studies in which experimental factors are always nested within others (called nested designs; see Chapter 1) cannot assess interactions among factors.

As can be imagined, one can consider higher level interactions. For example, in a three-factor factorial study with factors A, B, and C, the three-factor interaction is denoted by A × B × C. As the number of factors grows, the practical interpretation of such high level interactions becomes difficult.

5.1.3 Advantages and Disadvantages of the Completely Randomized Design

As has been illustrated by the various examples given above, the completely randomized design is relatively simple to administer as experimental units are simply randomized to each of the various treatments or treatment combinations. Because each unit receives only a single treatment or combination, there is potentially a shorter time required per subject than there would be with other possible designs (see below). Hence, for example, lower dropout rates would be expected in clinical studies using completely randomized designs.

However, the obvious disadvantage of completely randomized designs is that comparisons among treatments must be made between subjects rather than within subjects. As subject-to-subject variation is often larger than within-subject variation, a larger number of subjects are often required than with other possible designs.

5.2 Stratified Design/Randomized Block Design

Blocks or strata are collections of like experimental units which are expected to respond similarly as a result of treatment. In a randomized block design, all treatments are applied within each block and treatments are compared within the blocks. The randomization procedure randomizes separately within each block (stratum). When a study is designed so that the number of experimental units within each block administered each treatment is equal, the design is called a balanced complete block design.

There are two major reasons for utilizing a randomized block design rather than a completely randomized design. Suppose that we can identify an important prognostic factor related to treatment outcome. Would it not be unfortunate if our randomization yielded a design which was greatly imbalanced on this factor? If this were the case, it would certainly be more difficult to distinguish the effects of treatment from the inherent differences due to this prognostic factor. For example, suppose that in a proposed clinical study, it is well known that stage of disease may be a good predictor of treatment success, irrespective of the treatment being administered. We could depend on randomization to automatically balance on disease stage. However, it would generally be most prudent to balance treatments across the various levels of this important variable.

As discussed in Sections 3.3 and 3.7, although a very large imbalance is unlikely under randomization, especially in a large study, there is a potential bias introduced by allowing this possibility. If it is feasible, why not eliminate this possibility beforehand by considering each block or stratum separately and randomly assigning treatments in a balanced manner within each stratum?

As illustrated earlier with the simple numerical example in Table 3.3, use of the randomized block design not only has the advantage of minimizing potential bias by automatically balancing on the blocking variable, it can considerably reduce the residual or unexplained variation in a study and hence make the estimation more precise.

In many instances the blocking factors are not of any interest at all, while in other studies we may want to learn how the treatments differ within particular blocks (or strata), and so the idea of follow-up analyses by block or stratum is important. Perhaps the major distinction in this regard is whether the blocking factor can be considered as a *fixed effect* or as a *random effect*.* A fixed effect is typically one with a small number of levels and the selection of the levels under study is by choice rather than by random selection. For example, in a clinical study one might want to randomize separately for males and females, or on the basis of a range of ages (e.g., by decade). At the end of the study, one could not only analyze the data overall, but also investigate whether the effects seen in the entire study were consistent within strata. Needless to say, if the outcomes in one or more strata were particularly important, then the sample size estimation should take this into account. That is, if one wanted to be able to draw conclusions at the stratum level as well as at the overall study level, then the study would need to be considerably larger.

Random effects are factors such that samples are supposedly drawn at random from a (theoretically) infinite population of such factors and the samples are therefore considered to be representative or typical of that population. We are not really interested in the random effects themselves or differences among them and we can consider their effects as nuisances to be dealt with in the statistical analysis. For example, suppose that only a limited number of units can be tested in a given time period. If we ensure that all treatment conditions are tested with the same number of experimental units within each time period, then the effect of period will balance out and the comparisons of interest, i.e., among treatment conditions, can be made without any difficulty.

---

* Many statisticians would choose to define cross-classified designs with all effects fixed simply as factorial studies. However, these studies would differ from randomized block designs only with respect to analyses for some parameters, but not with respect to study design.

(See Figure 3.1 and discussion as an example in which assays are blocked by day and period within day.) As a second example, consider the case where multiple treatments can be administered to the same subject, either simultaneously, or in time order. Again, we are not specifically interested in a "subject effect" only on how treatments differ within each subject.

There are many special cases of randomized block designs depending on the issue of how a block is defined. When a block consists of a subject who can receive all treatments, the order of presentation of the treatments is randomized for each subject. In the special case of two treatments and blocks consisting of exactly two like subjects, the design is called a *matched pair design*.

Several examples will help to illustrate the concepts involved in randomized block designs and their application in biomedical research.

Example 5.3 (Continuation of Example 3.5)

In a clinical study conducted in the U.K. to assess the efficacy and safety of a potential new treatment for migraines, subjects were first stratified on the basis of average number of migraine headaches per month for the previous 3-month period. Two separate strata were defined in the study:

1. Patients with at least two, but no more than four migraines per month were entered into the "low migraine" stratum.
2. Patients with more than four migraines per month were entered into the "high migraine" stratum.

The primary motivation for stratification (blocking) by prior frequency of migraines was to ensure comparability between treatments (active and placebo) on this important prognostic factor. A second motivation was to be able to efficiently assess treatment effects in both high and low migraine patients.

Example 5.4 (Continuation of Example 3.6)

In a veterinary study in anemic dogs to assess the efficacy and safety of a temporary oxygen carrying blood replacement, patients were first stratified on the basis of cause of anemia (blood loss, hemolysis,* or ineffective erythropoiesis**) and then randomized to either active treatment or control within each stratum and investigational site in this multicenter study. Again, the primary motivation for stratification was

---

* Hemolysis is the disruption of the integrity of the red cell membrane causing release of hemoglobin.
** Erythropoiesis is the production of erythrocytes.

to balance treatment allocations on this important prognostic indicator. The secondary purpose was to evaluate efficacy for each cause of anemia.

Although this veterinary clinical study clearly demonstrated the superiority of the treatment to the control condition, it was instructive to assess its effect by stratum. Two of the major efficacy endpoints defined in the study protocol were whether each case was classified as a success or failure in a 24 h period, and the time to failure. Statistically significant differences were clearly demonstrated for each of these two endpoints. In addition, a clear indication of treatment efficacy was observed for these endpoints in each stratum of the study. In addition, we observed that success rates were generally lower and time to failure shorter in the stratum where the cause of anemia was identified as hemolysis.

Example 5.5

In a clinical study designed to assess the skin irritation potential of a new dermatological treatment, all five treatments in the study were administered simultaneously on the back of each normal volunteer. The five treatments (vehicle control, low treatment dose, intermediate treatment dose, high treatment dose, and positive control) were administered in small uniform samples on five separate areas (called A, B, C, D, E) on the back of each volunteer. The primary motivation for the design was the recognition that people vary considerably in their innate sensitivities, so that within-subject comparisons would be considerably more precise than between-subject comparisons, and so the total sample size (number of volunteers) in the study could be considerably lower than if each volunteer was administered only a single compound. The overall sensitivity of individual subjects was not of interest.

As illustrated, the advantage of the randomized block design over a completely randomized design is that comparisons are made within blocks, which are like units expected to respond similarly. If the blocks are the subjects themselves, than the comparisons may be more precise, i.e., less variable, than comparisons among subjects. Hence, lower sample sizes would generally be required. This is generally the case as repeated measurements within a subject tend to be positively correlated (similar).

In considering disadvantages of the randomized block design, suppose that each of T treatments is administered sequentially in time to each subject. Then the total duration of the study must be at least T periods. Worse yet, if there is a "washout" period between each pair of treatment periods, as is often the case, then the duration of the study extends for (T-1) more washout periods for each subject. In addition to the potential for an extremely long study if each treatment period is prolonged, there is also the heightened chance of dropouts in a study such as this as each subject's involvement is considerably lengthened. Missing data due to dropouts are then a cause of bias and much concern.

5.3   Crossover Study

Crossover studies are special cases of randomized block designs in which experimental subjects constitute the blocks and treatments are administered sequentially in time. The order of treatments is balanced across all subjects, i.e., each subject is administered several treatments, with studies generally planned to have each sequence of treatments given the same number of times.

Crossover designs can be used within many contexts of biomedical research (clinical and preclinical studies as well as engineering studies) as long as the outcome is nondestructive and the study itself does not take place over such an extended period of time that the early results and the late results cannot really be considered comparable for each subject.

Use of a crossover is illustrated with the simplest possible such design, the two period crossover design as depicted in Figure 5.2. Subjects are randomized to one of two sequences, either Sequence 1 or Sequence 2, usually with an equal number of subjects per sequence. Subjects in Sequence 1 are administered Treatment A in the first period of the study followed by Treatment B in the second period. Subjects in Sequence 2 are administered the treatments in the opposite order.

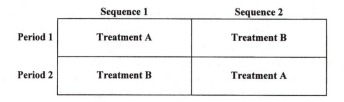

	Sequence 1	Sequence 2
Period 1	Treatment A	Treatment B
Period 2	Treatment B	Treatment A

Figure 5.2 Two period crossover design.

Unlike the design in which each subject is his/her own control (Section 5.5) and in which the control treatment always precedes the active treatment, in the balanced two period crossover, an equal number of subjects receive each treatment first. Thus, there is no confounding of treatments with time order and it is easy to separate their individual effects.

There is often a washout period between Period 1 and Period 2. In bioequivalence studies, for example, the length of a washout period is often at least five half-lives of the drug under study.* In addition, one can also include baseline measurements (measurements taken just prior to treatment) at the start of Period 1 and Period 2.

Example 5.6

As an illustration of the two period crossover design, a common use of this design is for the assessment of the availability of a new formulation of a drug, e.g., an orally administered form vs. intravenous administration. Normal volunteers in the sequence 1 group are administered the drug intravenously in Period 1 and in the oral formulation in Period 2. The same number of subjects are assigned to sequence 2 and are given the formulations in reverse order. The primary goal of the study is to assess the rate and extent of absorption of the oral formulation relative to the IV formulation by comparing pharmacokinetic parameters such as the area under the blood-concentration time curve (AUC) and maximum concentration (C_{max}).

Example 5.7

As the number of treatments increases, so does the number of periods. For example, Figure 5.3 illustrates the possible sequences in a three period crossover design. Notice that the number of possible sequences of treatment administration is now six. One way to see this is to remember that each subject is exposed to a different treatment in each period. Thus, the number of combinations in the first period is three, the number in the second period is two (one has already been chosen for period one), and there is only one remaining for the third period. The total number of sequences is therefore $3 \times 2 \times 1 = 6$. In Figure 5.3, these six sequences have been displayed in two separate squares (Square 1 and Square 2). Each square constitutes by itself what is termed a *Latin Square* design as each treatment appears exactly once in each row and column of the square. While it would not be necessary to use both squares in a three period crossover study, it would certainly be desirable to do so, especially if the number of subjects in the study could be made a multiple of six (so that the design could be nicely balanced). In this case, all possible sequences would be represented in the study.

---

* The elimination phase of the blood concentration vs. time curve in a bioequivalence study is usually characterized by "exponential decay." Thus, if one were to plot log concentration vs. time (a semi-log plot), the data would be approximately linear. The time interval required for the concentration to decrease by one half is called the half-life of the drug.

Square 1

	Sequence 1	Sequence 2	Sequence 3
Period 1	Treatment A	Treatment B	Treatment C
Period 2	Treatment B	Treatment C	Treatment A
Period 3	Treatment C	Treatment A	Treatment B

Square 2

	Sequence 1	Sequence 2	Sequence 3
Period 1	Treatment A	Treatment C	Treatment B
Period 2	Treatment C	Treatment B	Treatment A
Period 3	Treatment B	Treatment A	Treatment C

Figure 5.3 Three period crossover design.

In addition, the use of both squares provides protection against the effects of treatment which may carry over from one period to the next and hence, allows for unbiased estimation of the direct effect of each treatment.

Extensions to four or more treatments/periods follow along the same lines. However, instances of crossover designs with many treatments are very rare in biomedical research as the number of periods required for each subject, and hence the total time involvement becomes prohibitive.

There are several variations to crossover designs which one also encounters from time to time. When one of the main concerns relates to carry-over effects, then designs in which each treatment is preceded by all others equally often are sometimes used to estimate the magnitude of this residual effect. These designs are called crossover designs balanced for residual effects. For example, in the clinical study of anti-depressant drugs, such drugs are known to have effects even after the patient stops taking them.

Another variation on the standard crossover study is when one or more of the treatments are administered more than once to a given subject. For example, in a study with two treatments, A and B, one could use a three period design in which the two sequences were A,B,A and B,A,B. Such designs are called *switch-back* designs. The goal of such a design might be to provide a more precise estimate of treatment and carry-over effects with a minimum number of subjects.

Another variation on the standard crossover is when there are fewer periods (and hence treatments per subject) than treatments. For example, in a study with four treatments (A, B, C, and D), subjects are administered two of the four treatments in such a way so that all treatments are equally represented in Period 1 and in Period 2, and all pairs of treatment combinations are also equally replicated. With this design, there are 12 possible sequences: AB, AC, AD, BC, BD, CD, and the same pairs again, but in reverse order. Designs which have the requisite balance as described above are termed *balanced incomplete block designs*. (See Westlake, 1973 for a discussion of balanced incomplete block designs for bioequivalence studies.)

In general, the main advantage of a crossover design relative to a completely randomized design is that comparisons are made within subjects rather than between subjects. For example, in the bioequivalence study described above, we can estimate a relative availability for each subject. Because such comparisons are often more precise than between-subject comparisons, fewer subjects may be required in such a study. Also, because treatment administration is balanced across subjects and time periods, we do not have the confounding that we do in other designs.

There are two potential disadvantages of crossover designs. First, there is the possibility of carry-over from one treatment to the next period. If the carry-over differs from treatment to treatment, then bias is introduced. This can be a very serious problem. Second, crossover studies require multiple treatment administrations (and perhaps also multiple washout periods) per subject, and therefore, they may take more time to complete. It may be difficult for many subjects to complete all periods. Subjects who drop out of the study before completing the last period cause the data to be unbalanced and make the statistical analyses more difficult.

5.4 Split Plot Design

5.4.1 Origin

The term *split plot* has its origin in agricultural research (e.g., see Yates, 1935). This type of design incorporates subsampling and allows one to make comparisons among different treatments at two or more sampling levels. In agricultural studies, fields are randomly assigned to different levels of a primary factor. In addition, smaller areas within the fields are also randomly assigned to one level of another secondary factor. Although the larger sampling unit, the field or plot, is assigned throughout to one level of the primary factor, secondary sampling units, referred to as the subplots, are assigned different levels of the secondary factor. The idea is that the fields themselves are so large that comparisons from field to field are not necessarily very precise. However, within a field, areas are more alike so that comparisons within each field will have higher power to detect differences among different treatment conditions, in this case, levels of the secondary factor.

Depending on the size of the field, further levels of subsampling (e.g., sub-subplots) can also be considered. While the split plot design is less common in biological research, it is sometimes employed, especially when the second factor relates to time or experimental order.

The two factor split plot design is depicted in Figure 5.4. Different levels of the first factor (A) are applied to experimental units at the plot level. The second factor (B) is applied at the subplot level. In the example shown, A has three levels (A1, A2, and A3) and B has two levels (B1 and B2). Each column in the figure identifies a plot (six plots are shown in the figure). Each cell identifies a subplot. For example, we see that in the first plot, the treatment A is administered at the third level (A3). The first subplot also gets level 1 of the treatment B (B1), whereas the second subplot gets the second level of treatment B (B2).

REPLICATION 1			REPLICATION 2		
A3	A1	A2	A1	A3	A2
B1	B2	B2	B1	B1	B2
B2	B1	B1	B2	B2	B1

Figure 5.4   Example of a split plot design.

5.4.2   Repeated Measures Design

Repeated measures designs are similar to split plot designs. In fact, if one of the factors being studied is time and each plot consists of a single subject, then the design is called a repeated measures design.* Such repeated measures designs are very common in biological research. In a repeated measures design, we are not only interested in assessing the overall effect of treatment, but also the effect of time, and the treatment by time interaction. Because we make multiple observations on the same subject over time, observations within a subject are correlated and the data analysis needs to take into account the correlation structure of the data. Needless to say, even with a repeated measures design, the second factor can introduce a real treatment condition other than time (e.g., dose level), but the analysis would proceed in the same way.

---

* Unlike the example in Figure 5.4, the levels of time cannot be randomly assigned.

5.4.3 Summary

The main advantage of a split plot design (in a two factor study) is that comparisons of the second factor (called B above) are made at the subplot level rather than at the plot level. If subplots are more alike than different plots, then such comparisons may be more precise. As can be imagined, however, the split plot design is not always applicable. For example, it would be difficult to implement when there are many factors under study. Similarly, it is clear that the classification into plots and subplots must be a natural one (recall the study design earlier with animals, lobes of the liver, fields, cells). The same considerations apply to repeated measures designs. Overall comparisons of the average response across all time points for treatments are made between subjects. Conversely, comparison of the overall time effect and treatment by time interactions are made within subjects and are generally more precise.

5.5 Types of Control

In many instances in which we are designing biomedical studies, there will be a standard treatment condition against which all others may be compared. Such a treatment condition is often referred to as a *reference standard* or a *control*. As such, the control can either be a negative control or an active (positive) control. Some studies, e.g., *in vitro* studies to investigate the genotoxicity of a particular chemical, typically employ both negative and positive controls simultaneously. Very often, the inclusion of a control group or treatment condition is absolutely essential to deriving a valid inference from an experimental study.

The purpose of such a control group (or condition) is to provide a background against which all other treatment conditions can be assessed. These control groups or conditions are run concurrently with the treatments of interest, so that the term *concurrent control* is often used to distinguish from control information available from historical data (*historical controls*). While information from historical controls is often useful in helping to put the results of the current experiment in perspective, data collected from concurrent controls provide the best basis of comparison. The reason for this lies in the possibility of bias being introduced due to variations in the experimental protocol or techniques between the historical study (studies) and the current one, time trends from experiment to experiment, genetic drift, different sources of experimental units, laboratory reagents, etc.

Even within the realm of concurrent controls, there are several choices about their form. The best choice for a given study will depend on the primary objectives of the study, the feasibility of negative and/or positive controls, ethical considerations in clinical trials, and issues related to sample sizes and costs. In the remainder of Section 5.5, we consider several types of controls, the appropriateness of each, and their advantages and disadvantages.

5.5.1   Negative Control

A negative control is any treatment condition which would be expected to produce a response at the same level as normal or background for the population of experimental units under study. As a result, the negative control provides results at the *nominal level* so that substantial changes from this level can be assessed for experimental treatments in the study.

5.5.1.1   Subject as His/Her Own Control

A very common design in early clinical trials (Phases I and II) is to utilize a subject as his or her own control. In this design, the subject is first evaluated under standard conditions (i.e., without treatment). The treatment is then applied, and the response is remeasured during or after treatment. Because the response is evaluated in each subject prior to treatment and also after initiation of treatment, this design has the property that all comparisons are made within rather than between subjects. Because intrasubject variability is often lower than intersubject variability, this design may require fewer subjects than designs in which comparisons are made between subjects.

There are, however, a number of strong assumptions being made when this design is utilized. As each untreated evaluation is made prior to each treated evaluation, the effect of treatment is totally confounded with treatment order (time). This is a potential source of bias. In order for a design like this to be valid, we must be strongly convinced (and perhaps be able to demonstrate to our critics) that the process under study is stable, at least for the duration of the study. A second difficulty will be that such a study is not blinded, neither to the investigators, nor to the subjects themselves. We have already discussed (Section 3.3) the issue of blinding for investigators/raters and subjects and its effect on minimization of bias. Therefore, for these reasons, studies which provide an additional condition (i.e., active or placebo control) and blinding as to condition are usually preferable.

As these are very likely large sources of experimental bias, it seems clear that this is not the design of choice under most situations. However, this design has some utility, especially in early stages of a clinical or research program, where a preliminary assessment of efficacy and safety (even if somewhat biased) may be useful in deciding whether to proceed with further, more complicated and elaborate studies. Despite this, one should keep in mind the shortcomings of using the subject as his/her own control and view experimental results obtained with such a design accordingly.

5.5.1.2 Untreated Control

The use of an untreated control therapy can be made in one of several ways: within the context of a completely randomized or randomized block design, or within the context of a crossover design. The major distinction is whether comparisons among treatments are made between or within subjects (experimental units). In either case, the untreated control condition is evaluated without any particular treatment being applied.

As an example of an untreated control group in the context of a randomized block design, we consider the clinical trial of a temporary oxygen carrying blood replacement in dogs being treated for anemia. Dogs were first stratified on the basis of cause of anemia, and then randomized to either the untreated control or active treatment group within each stratum.

Like the use of subjects as his/her own control, subjects are aware of whether they are receiving active therapy or not, and as a result, such a study cannot be blinded to the subjects. Similarly, investigators are often aware of the treatment status of subjects and cannot always be blinded either. Under most conditions, raters can be blinded to the treatment status of the subjects, as was the case in the clinical study mentioned above.

When an untreated control condition is utilized as part of a crossover study (see Section 5.3), lack of treatment is incorporated as one of the "treatments" under study. This design has the advantage that comparisons are made within subjects rather than between subjects, but still suffers from the problems of lack of blinding as discussed above.

5.5.1.3 Placebo Control

The final type of negative control to be discussed is the placebo control. By a placebo control we mean a treatment which matches the active treatment(s), but without the active ingredient. Several examples will help to illustrate how a placebo control is utilized.

Suppose that one is conducting a clinical study in which the test compound is being administered to patients in tablet form. A matching placebo tablet is formulated which is identical in size and appearance to the tablet containing the test compound. The placebo tablet should contain the same inert ingredients as the test tablet, and only differ by not having the active compound. Any packaging of placebo tablets should also be identical to that of the test compound, although all such packaging (e.g., vials with labels) should provide identifying information, usually in the form of code numbers. This coded information or "blinding" can be broken, if necessary, in the case of emergencies by contacting a particular individual specified in the study protocol.

Under other conditions, the term *vehicle control* is used to describe the matching treatment condition with the exception of the active ingredient or test compound under study. For example, in many toxicology studies, it is typical to include a vehicle control group in which animals are administered the solution or suspension in which the active compound would usually be given, but without the active compound. If treatment is to be administered by gavage, then animals in the vehicle control group are gavaged with the vehicle only.

Another term which is often used in this same context is *sham control*. In a study of the carcinogenicity of polyethylene implants, for example, where the surgical implantation of the test material is accomplished at the initiation of the study, a sham control group may be included. In the sham control group, animals undergo surgery, but without any implant.

What all these have in common is that the group is treated identically to active treatment, except for the test compound administration. Because active treatment and placebo control groups differ only with respect to presence or absence of the test compound, and because experimental units are randomized to groups (or treatment order if multiple treatments per subject), any differences in response can be attributed to the test compound. By making all conditions among groups (or among multiple treatment administrations) as similar as possible, one hopes to eliminate or at least minimize any bias that might be introduced by such dissimilarities. In particular, most single- and double-blinded studies are conducted using placebo control.

Although placebo control is always desirable, under some instances there may be ethical considerations which preclude its use. For example, in a long-term study in which active treatments are applied by injection, it is questionable as to whether it is necessary to expose control group patients to repeated painful injections for a sustained period. Instead, it may be more practical to utilize an untreated control group with blinding of investigators as to treatment group status.

5.5.2   Active Control

Whereas a negative control is a condition which is not expected to elicit a response different from background, an active or positive control is a condition which can be expected to elicit a substantial response in the type of experiment or test system under study. There are two different reasons for incorporating an active control in a study. The first is to validate the experiment or test system, i.e., to demonstrate that it is performing within the limits that are expected. The second is to serve as the basis of comparison when a negative control is inappropriate or infeasible.

In genotoxicity studies, for example, known mutagens are typically included as positive controls to validate a particular study. These compounds are known to produce mutant colonies at a very high rate. By including a positive control in such a study and verifying the expected results for that group, it allows one to have greater confidence in negative findings for test compounds.

Example 5.8 (Continuation of Example 5.5)

Another example of the use of a positive control was in the evaluation of the skin irritation potential of several dermatological compounds, a study briefly mentioned in Section 3.4. In fact, the five compounds under study included a vehicle control, a low dose of the test agent, an intermediate dose of the test agent, a high dose of the test agent, and a positive control. The major comparisons to be made were relative to vehicle. The questions of interest were

1. Does the test agent have irritant properties above its vehicle?
2. At what dose levels is this irritation present?

The purpose of the positive control (a known skin irritant) was to validate the test system.

The second use of a positive control in biomedical studies is to provide a basis of comparison. In some instances, a negative control is inappropriate. For example, in bioequivalence testing to compare the rate and extent of absorption of a new formulation of a drug product to a standard formulation (intravenous administration or marketed formulation), a negative control would not make sense. The standard formulation serves as the basis of comparison.

In some cases, a negative control may be medically unethical. If the condition under study is potentially life threatening or if permanent disability would result from lack of treatment, then withholding treatment would be wrong. In a comparative study under such conditions, only a positive control can be considered. The positive control would usually be the current standard therapy.

Positively controlled studies are usually more difficult to conduct than are negatively controlled studies. For one thing, sample sizes will often be considerably larger with the former than with the latter. Differences of interest between two active therapies (the standard and the test) will be considerably less. Moreover, standard methods of hypothesis testing are usually inappropriate for positively controlled studies. Presumably, the objective of such a study is to demonstrate equivalence or at most a small difference between therapies. Suppose that one simply considers the null hypothesis that the therapies are equal. Failure to reject could be only due to having low power for the statistical test. Alternatively, if a very precise study were performed, even a small (medically insignificant) difference could be detected as statistically significant. For these reasons, conventional hypothesis testing should not generally be employed in positive controlled studies, and hence, such studies should be designed to facilitate estimation, confidence interval calculation, or nonstandard tests of hypothesis (e.g., nonequivalence).* It seems that only if it is important to detect a small benefit of a new therapy (instead of equivalence) should standard hypothesis testing methods be used.

5.6   *Dose Selection in Dose-Response Studies*

In biomedical research, examples abound for studies in which one or more factors are studied in at least two different quantitative levels. In clinical trials, for example, perhaps the most common design is one in which there is a placebo treatment (a zero dose) and one or more levels of a test drug. In toxicology studies to assess the safety of compounds in animal models, results at several doses are often compared to those obtained in a vehicle control group (the zero dose). In engineering studies to assess the effects of various factors on such outcomes as process yield or reliability of an assay or production process, factors are often studied at multiple quantitative levels in order to optimize yield or reliability or any other important characteristics under study.

In such situations, which we will term dose-response studies, how many dose levels should be incorporated? How should doses be chosen?

The answers to such questions relate, as they should, to the study objectives. The study objectives should state in sufficient detail the purposes of the study and what features of the dose-response relationship are most important. It is not enough to say the objective of the study will be to "evaluate the response Y as a function of the dose X." In order to select the number and levels of doses for the study, one has to provide one or more specific objectives.

In order to clarify this, I will identify four possible purposes of dose-response studies:

1. To determine if the factor under study has any effect at all on the response
2. To estimate parameters of a particular known model in the population under study
3. To assess whether a particular dose-response model holds relative to likely alternative models
4. To select one or more possible models from a collection of candidate models.

---

* In the context of bioequivalence testing, many authors have argued against standard tests of hypotheses (see Metzler, 1974; Westlake, 1972; Selwyn et al., 1981; and Schuirmann, 1987).

These and perhaps other specific objectives help the study designer in identifying and evaluating good designs in terms of doses to be studied. I will illustrate these concepts with a practical example.

Example 5.9

Suppose that we are interested in testing whether a potential antihypertensive drug has an effect on (lowering) blood pressure. Suppose further that the dose-response for the drug, i.e., decrease in systolic or diastolic blood pressure versus dose, is linear in the range from 0 to the level A, as depicted in Figure 5.5. The most effective design for both testing for the effect of the drug and also for estimating the form of the dose-response (in this case the slope of the line connecting the two points in Figure 5.5) is to allocate one half of the data points at a zero dose and one half of the data points at the dose A. If the design is a completely randomized design, then half of the subjects are randomized to the placebo and half to the dose A. If the design is a two period crossover, then subjects would be administered placebo in one period and the drug at dose A in the other period. Again, half of the subjects would be randomized to the order (placebo, dose A), and the other half to the order (dose A, placebo). Most researchers would probably not utilize this selection of data points and would probably include one or more intermediate dose levels such as A/2 or A/3 and 2A/3. By doing so, they dilute their ability to detect a significant drug effect (i.e., the power of the test for slope) and decrease the precision with which the slope is estimated.

This is not to say, of course, that only two dose levels should necessarily be utilized in every study. The point I am trying to make is that, if the major objective of the study is to assess a drug effect and estimate its magnitude in a situation in which the response is likely to be linear or very close to linear, then the optimal design in terms of dose selection is a two-point design, with doses at the most extreme values possible (0 and A in the above example).

However, if it is unsure whether the dose-response is linear in a particular range, and in fact if the dose-response is likely to be non-monotonic* as depicted in Figure 5.6, then a two-point design, such as one with dose points at zero and A, will have very low power to detect any drug effect at all. In fact, if the true response at A is exactly equal

---

* A monotonic dose-response is one which is either nondecreasing (i.e., increases or remains the same, but never decreases) or nonincreasing (i.e., decreases or remains the same, but never increases) throughout the dose range studied. Exceptions to this would be called nonmonotonic dose-responses. Although such functions could wiggle up and down a few times, the more likely alternatives to monotonic dose-response curves would increase and then decrease (as in Figure 5.6) or decrease and then increase (the flip side of Figure 5.6).

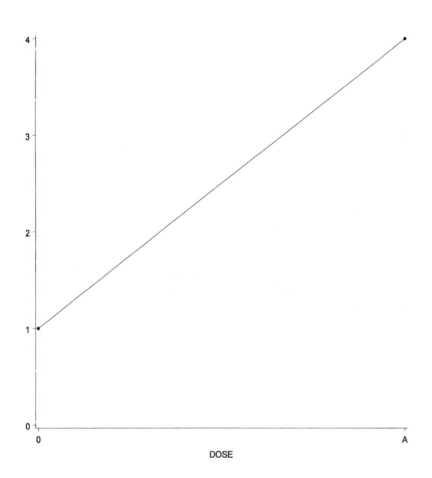

Figure 5.5 Example of linear dose-response.

to the response at zero, then a design such as this will have power only equal to the level of the statistical test (i.e., 1, 5, 10%, etc.). In fact, in situations in which the actual dose-response is quadratic, as depicted in the figure, the optimal dose-response design will have three dose levels (at the apex of the parabola, and at two dose points equidistant from it in both directions). Again, spacing of the doses should be as wide as possible within the range of feasible doses. In general, if the dose-response can be characterized as a low degree polynomial function, say of degree K, then the optimal dose design will have K+1 dose points.

What this extended example illustrates is that a satisfactory selection of dose levels in a dose-response study will depend critically on the objectives of the study and what is known about the dose-response.

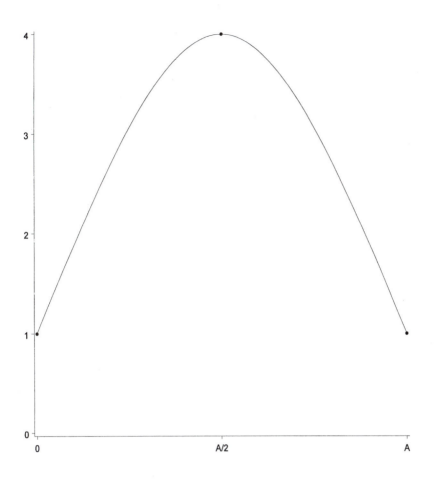

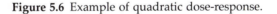

Figure 5.6 Example of quadratic dose-response.

If one of the purposes of the study is to test for a linear dose-response, where a quadratic or similar dose-response is a possible alternative, then at least three dose levels are required and the optimal selection of these doses will depend on the true response itself.

What can one do if little is known about the dose-response prior to a particular study? Here, it seems that a pilot study (see Section 3.2) would be particularly useful in identifying the likely range of doses in which there is an effect and also to possibly identify, or at least narrow down, a collection of candidate models for consideration in the selection of doses.

We have seen above how the progression from a strictly linear model, in which two dose levels provide the optimum design, to a quadratic model leads to an additional dose level. When would it be necessary to have more than three levels?

Although cubic, quartic, and even higher degree polynomial models are possible, perhaps more prevalent dose-responses of interest would be as depicted in Figure 5.7. Here we see a threshold in dose up to dose B, a linear response from dose B to dose C, and then a plateau beyond dose C. In order to adequately characterize the dose-response, it would be necessary to have design points (doses) at A, B, C, and D, so that four points would be required.

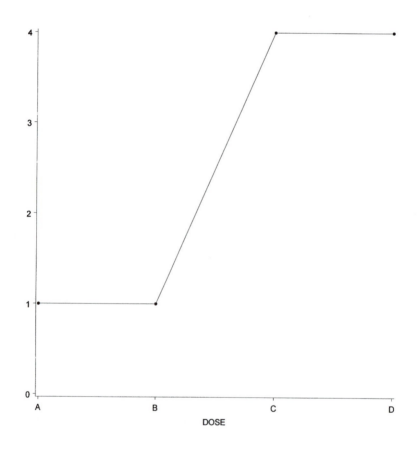

Figure 5.7

Assessing several alternative but similar models can be very difficult. Consider, for example, the sigmoid shaped curve depicted in Figure 5.8. This is very similar in overall appearance to the dose-response curve generated by the line segments in Figure 5.7. It would be very difficult to distinguish between these two models, especially in a study in which the underlying variation was large relative to the differences between the two curves. Would it be important to tell whether the piecewise linear dose-response in Figure 5.7 or the sigmoid curve in Figure 5.8 (or perhaps another model) was the correct dose-response curve in a given study? This depends completely on the objectives of the study and its context. If a major purpose of the study is simply to characterize the general shape of the dose response, then both models probably provide an adequate representation. For example, one could say that the response is flat or nearly flat in the low dose region, rises sharply in a linear fashion, and then plateaus at higher doses. If desired, one

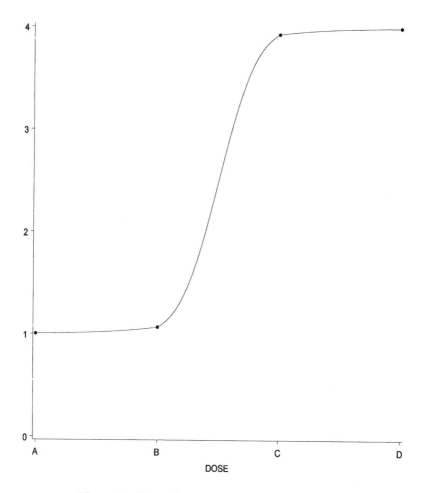

Figure 5.8 Example of "sigmoid" dose-response.

could provide both point and interval estimates of the predicted response at selected doses.

On the other hand, suppose that it was very important to know whether the true dose-response curve was the former or the latter of these two models. For example, existence of a true threshold (lower end) or a true plateau might have some important biological, economic, or policy consequences. In such a case, it would be beneficial to be able to design a study to distinguish between the two competing models. With such closely aligned functions, however, such a study would necessitate inclusion of at least four or five dose levels, each with considerable replication.

In summary, in a study which has as one of its objectives the assessment of the response as a function of dose, the selection of doses will depend critically on the specific objectives of the study as well as what is already known about the form of the dose-response function. I have listed above four possible objectives in a dose-response study, but there may be others. It is critical to identify the more important goals in each study and to rank them. It is clear from the above examples that selection of an optimal or even an appropriate design will depend on these factors. Distinguishing between similar dose-response models is a difficult task and can only be accomplished with large studies.

5.7 Multicenter Studies

Starting with Phase II, most clinical studies are multicenter studies. There are several reasons for this. First, studies in Phase II and Phase III are typically larger than Phase I studies. It would be difficult, if not impossible, for a single center to recruit all the patients that were required in a study. Even if it were possible, the time involved would make the study much, much longer than a multicenter study. A second reason for conducting such studies in several centers is to be able to demonstrate that results are repeatable from site to site and are not confined to a single hospital or clinic.

Another frequently encountered multicenter study is for assay validation of a chemical or biological assay. Such a project may have as its objectives to demonstrate that the assay is sufficiently accurate, precise, linear over a particular concentration range, and repeatable from laboratory to laboratory. Because laboratories will vary with regard to their environmental conditions, personnel, reagents, etc., it is important to show that assay results can be replicated among laboratories. The laboratory to laboratory variation is one component of the total variation in the assay. In a well-designed interlaboratory study, the proportion of the variation due to this component and all others can be estimated.

It is rare to have preclinical laboratory studies conducted in multiple sites simultaneously. Perhaps the closest comparison would be when a chronic or carcinogenicity study in laboratory animals is conducted in multiple animal rooms within the same facility. The identical design issues that one faces in multicenter studies are also faced in a study of this kind.

I have already discussed some of the issues related to multicenter clinical trials, i.e., generation and acceptance of a common written protocol and the importance of a kickoff meeting for investigators to iron out potential differences prior to the start of a study (Section 3.3). Otherwise, the same general experimental design principles as always apply in the multicenter situation. The effect of having multiple sites, however, is to introduce another level of blocking or stratification. For example, suppose that a multicenter clinical trial is conceived as a two-factor factorial study with factors A and B. One not only has to consider these factors, but also the investigational site or center as an additional factor in the study. Like the randomized block design discussed earlier, it certainly makes sense to randomize patients in each site separately and to ensure that all treatment combinations are represented at every site. In fact, the ideal allocation would be to have an equal number of patients randomized to each treatment condition at every site.

In most clinical studies, patient recruitment occurs over time, and at any given time point the number of patients recruited at each center will be different. While it would be best and simplest if each center would ultimately end up with the same number of completed patients (and as long as we are developing a wish list, we should not have any dropouts), this is almost never the case and it is likely that sample sizes will be unbalanced among the centers. Some centers will treat a large number of patients and others very few.

Despite this, our goal as study designers is to achieve as much balance and equal representation as possible. Suppose for example, that A and B are each being studied at two levels so that the total number of treatment combinations would be four. My recommendation would be to randomize each center separately with patients randomized in sets of eight (two times the number of treatment combinations), with two patients assigned to each treatment combination in each such set. Usually, the sample size is determined based on the totality of patients from all centers, so that when this sample size is achieved, the study will be terminated. Hence, treatment allocations will not necessarily be equal at all centers. However, by ensuring balance after every set of eight patients at each center, approximate balance will be achieved over all centers. Moreover, by balancing every eight patients rather than after every four, it may make the blinding slightly more difficult to break for an extremely keen and curious investigator.

As can easily be seen, the above example simply constitutes a randomized block design where the centers are considered as blocks. As another example, suppose that the basic design is itself a randomized block design. Again, we could view each combination of center and block as a newly defined block and again randomize to treatments within these new blocks.

Some of the studies mentioned as examples earlier were in fact multicenter studies and each center was randomized separately. In the migraine study, for example, each center was randomized separately and individual randomization schedules were prepared for each center/stratum combination. The veterinary study of the oxygen-carrying blood replacement was a multicenter study as well and patients were separately randomized to active treatment or control in each combination of center and stratum (cause of anemia).

Thus, experimental designs for multicenter studies pose no additional problems other than those that have already been discussed. The factor of center or site adds an additional level of stratification or blocking, so that each center should be randomized separately. In experiments in which subjects are entered over time, it is convenient to employ randomization schemes that routinely balance among treatments for each center.

5.8   Summary

The overriding theme of this chapter has been that there is a great variety of experimental designs which can be considered for many biomedical studies. Most designs will have conditions under which they are applicable, as well as their own advantages and disadvantages as indicated in Table 5.5. In planning a new study, one should examine each possible design, evaluate its relative advantages and shortcomings, and select the one that appears to provide the best features for the particular study at hand.

Part of the discussion in Sections 5.5 and 5.6 concerned the number and types of treatment groups in a study. In general, the number of groups should be kept to a minimum. The indiscriminate addition of extra groups or treatment conditions should always be avoided as this tends to dilute the effects of real differences seen in only a single or a few groups in the study.

Table 5.5 Characteristics of some common experimental designs

Design (description and conditions for use)	Advantages	Disadvantages
Completely randomized design: Subjects are randomized to treatment groups or conditions in such a way that each group or condition is equally likely[a]	Simple to administer, one treatment period per subject, no dropout problem	Comparison of treatments between rather than within subjects
Stratified design/ randomized block design: Subjects are, randomized to treatments or conditions separately within each block or stratum	Comparisons made within blocks which are "like units," often smaller sample sizes are required than for a completely randomized design	Extended duration for each subject if treatments are administered in time, dropouts are a problem
Crossover design: Subjects are randomized to sequences, where each sequence contains all treatment conditions	Comparisons made within subjects rather than between subjects, often smaller sample sizes are required than for a completely randomized design	Potential for carryover from treatments into subsequent periods. Extended duration for each subject, dropouts are a problem
Split plot design: The first experimental factor is randomly allocated at the highest level of sampling (plots). Subsequent factors are randomly allocated to lower levels of sampling (subplots and so on)	Comparisons of secondary factors are made at lower levels of sampling which may be more precise than the primary level	Only applicable in special situations where the classification into "plots" and "subplots" is a natural one

[a] Absent other considerations, I would almost always advocate equal sample sizes for all treatment conditions for this design and all other designs as well.

chapter six

Sequential Clinical Trials

Who would not like to be able to terminate a study early and still be able to reach a sound conclusion? Consider the time and resources that can be saved by doing this. Further, if the final result of a clinical (or other study) is that a new promising therapy can finally be applied in a large patient population, think of the advantage of being able to bring this product to market early. This is exactly the motivation for sequential trials. I have titled this chapter "Sequential Clinical Trials" only because the application of sequential experimentation is perhaps more prevalent with respect to clinical trials than it is to nonclinical and/or engineering studies. This is not to say, of course, that the methods and strategies discussed herein cannot be used in other arenas. In fact, because of its wide applicability, the topic of sequential experimentation, as well as the broader topic of experimental design, is very important in all scientific research.

Many scientists are probably not familiar with the idea of sequential experimentation in a formal sense. I provide some background by describing the history of this topic, especially as it relates to clinical studies.

6.1 History

The origin of sequential testing occurred during World War II and was motivated in part by the need to reach decisions quickly and with a minimum of resources expended. Many of the theoretical and practical aspects of sequential testing were developed by the Statistical Research Group (SRG) at Columbia University. The statistician Abraham Wald is regarded by many as the founder of sequential statistical analysis (Wald, 1947).

In applying statistical concepts to ordnance testing (circa 1943), it was realized that very large sample sizes were required to satisfy standard statistical criteria, whereas experienced Navy personnel who were performing the actual tests would realize considerably earlier whether the experiment was a success or a failure. Upon reflecting on this process, statisticians at the SRG realized (Wallis, 1980)*:

> ... that it might pay to design a test in order to capitalize on this sequential feature; that is, it might pay to use a test which would not be as efficient as the classical tests if a sample of exactly *N* were to be taken, but which would more than often offset this disadvantage by providing a good chance of terminating early when used sequentially.

Hence the era of sequential experimentation began. The history of sequential clinical trials can easily be traced back to the early 1950s (e.g., see Armitage, 1954). Because of their potentially great impact on both sample size and trial duration, there has been considerable interest in and development of statistical methods for sequential clinical trials in the intervening period, and much of this research continues even today. It would be difficult to provide a complete bibliography of the statistical and medical literature on sequential trials. An interested reader might refer to the review article by Jennison and Turnbull (1990) or the recent book by Whitehead (1992; for the more mathematically inclined).

Over the years, there have been many ideas and refinements that have been incorporated into the design and analysis of sequential trials. Wald's original approach, applicable to the clinical setting as well as to a great variety of other applications, was termed a *sequential probability ratio test* (SPRT) and has the property that it minimizes the expected sample size when either the null or (single point) alternative hypothesis holds.

Other statisticians have developed competing methods which have different optimality properties. Repeated tests of significance (e.g., see Armitage, 1975) allow for the repeated testing of the null hypothesis throughout the duration of the study on a regular, frequent basis. Each test is performed at a very low level (e.g., 0.003 to 0.007 with between 20 and 200 possible interim analyses for a nominal two-sided test at the 0.05 level; see Geller and Pocock, 1987). The idea is that even with this many planned analyses, the overall false positive rate will still be satisfactory because of the very low false positive rate for each test and the correlation among the repeated tests (as they depend on the same accumulating data).

Another approach to sequential experimentation has been to use *group sequential tests* (Pocock, 1977; Geller and Pocock, 1987). These group sequential tests allow for only a small number of interim analyses, usually two to five, and provide much of the benefit of early stopping as do completely sequential studies, and may be easier to implement in practice, especially for a multicenter clinical trial.

The original SPRT developed by Wald has the slightly disturbing property that it can, in theory, go on indefinitely.* Such a procedure is called an open-ended test. In contrast, Whitehead (1983, 1992) has developed an alternative closed test called the *triangular test* which is a sequential test with finite maximum sample size. This test is not as efficient early on as the SPRT, but has tighter boundaries later in the trial, and also performs well when the true alternative is part way between the null and postulated alternative hypotheses.

There are other approaches to sequential clinical testing. The so-called "play the winner rule" and like strategies (e.g., see Zelen, 1969) assign patients to a better-performing therapy (usually with random assignments based on probabilities calculated from the accumulating data) rather than equally. Such procedures have as their goals not only to obtain information about competing therapies, but also to allocate more patients to superior therapies during the trial. Another approach to sequential testing incorporates a reestimation of sample size after some data have been collected (Shih, 1992).

6.2   Rationale

Suppose that we are conducting a clinical study to compare a new promising therapy to a standard one. We hope that the new therapy will prove superior to the standard therapy and thereby offer considerable benefits to the patient population under study. If we employ a fixed-size design for which the number of patients to be evaluated in the study is determined in advance, as indicated in Chapter 4, then we really do not know the results of the study until the last patient is evaluated. However, in most clinical trials, patient recruitment accrues over a period of weeks, months, or even years, so that final data on some proportion of the patients may be available well before the last patients are recruited. Suppose that we have access to these data and it is clear from preliminary results that the new therapy is exceedingly superior to the standard therapy, perhaps even by a considerable magnitude over what was anticipated during study planning and what was used to generate the sample size for the study. Would it not be desirable to be able to terminate the study at this point or shortly thereafter and conclude that the new therapy is superior?

---

* The probability of extremely large sample sizes, however, is very small.

Conversely, suppose that we are again in the same situation, but in fact, the new therapy is found to be inferior to the standard therapy in terms of its efficacy. We have enough data so that it is unlikely that this preliminary trend would be reversed for the remainder of the patients to be entered into the trial. Would it not be prudent to be able to terminate the trial at an early stage with the conclusion that the new therapy could not be shown to be superior to the standard? A slight variation on this theme would be if the new therapy were about as effective as the standard one, but its side effects were worse. In this case, we would want to conclude that a significant benefit could not be claimed for the new therapy.

The difficulty in being able to do this, as sensible as it seems, is that in general, the protocol specifies in advance what the methods for statistical data analysis will be. If the major approach is a test of a null hypothesis of equivalence of the new and standard therapies, say at the 0.05 level (i.e., the false positive rate is set at 5%), and one conducts more than one test at the 0.05 level, then the overall false positive rate will be inflated above this nominal level. The more tests that one does, the worse the situation becomes.

In order to maintain an overall false positive rate as specified, one needs to take into account that there will be several "looks" at the data, or what are called interim analyses. The actual analysis phase of a study of this kind is called a sequential analysis and the design aspects are referred to as a sequential design.

Can one simply consider the sequential aspect of the study to be an analysis issue and not part of the study design? No, because these aspects of the study must be incorporated in the study protocol if the study can be terminated early on the basis of the interim analyses. The specific rule used to determine when a study can be terminated (stopped) is called a stopping rule.

6.3   Sequential Designs

In a study employing a sequential design, experimental data are evaluated frequently during the trial to determine if a conclusion can be reached based on the data available at that time. One of the study endpoints must be identified as the primary variable in these interim analyses and used as the basis for decision making. In addition, the form of the stopping rule must be specified in advance.

Like the sample size estimation problem for hypothesis testing, the key attributes which determine the design are the false positive rate, false negative rate (or equivalently the expected power of the test), the null hypothesis to be tested, and the alternative hypothesis of interest. Data are accumulated over time. At each instance at which an interim analysis is performed, there are three possible outcomes: (1) the null hypothesis is rejected and it is concluded that the alternative hypothesis holds; (2) the null hypothesis is accepted; or (3) it is determined that there is insufficient information to make a decision so that the study continues. Notice that this approach differs from a fixed sample design for which only the first two options above are possible at the end of the study.

The sequential design is then completely specified by identifying a test statistic and the form of the stopping rule. As indicated above, different stopping rules will have slightly different optimality properties depending on their specific form, but the general rule is the same.

Example 6.1

Suppose for example that we are studying a process where the primary endpoint is dichotomous, say either a success or a failure. We are comparing a new therapy to a standard therapy and want to determine if the new therapy is superior (alternative hypothesis) to the standard or no better than the standard (null hypothesis). We could generate a randomization schedule in which patients are randomized to either the new therapy or the standard therapy in such a way that over the long run (and ideally over the short run too), approximately half of the patients would be randomized to each therapy. Alternatively, patients might be matched according to baseline variables and/or prognostic factors and entered into the study as matched pairs (see Section 5.2). If the determination of success or failure can be made quickly relative to patient accrual in the study, then a sequential design might be suitable in such a situation. With either randomization scheme above, the form of the test statistic will be related to the difference in success rates between the new therapy and the standard therapy. Suppose for example, that the standard therapy has historically had a success rate of 35%. We postulate that a clinically relevant improvement would be if the new therapy had a success rate of 50% or more.

We collect data on success rates as patients enter the trial and their outcomes are determined. At each point at which an outcome (or pair of outcomes) is determined, we update the test statistic to assess whether we now have sufficient evidence to reach conclusions (1) or (2) above or whether the study should continue as in option (3). This process can be depicted graphically as in Figure 6.1 which illustrates the rejection region of the sequential test, the acceptance region of the sequential test, and the continuation region for the sequential test of the sequential probability ratio (SPRT) form. The horizontal axis represents accumulating information (V) in the form of increasing sample size. The vertical axis represents the test statistic (Z), which in this case is related to the difference in success rates between the new and standard therapies. For instance, if the experimental rate of the new therapy was 70% and the experimental rate of the standard therapy was 32%, this would characterize a situation in which the observed difference between therapies was even larger than postulated. If the sample size at this stage was sufficient to conclude that this was not just a chance finding, i.e., due to "sampling variation," then the null hypothesis would be rejected at this point.

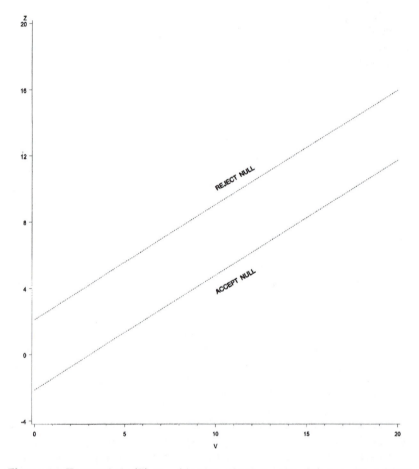

Figure 6.1 Test statistic (Z) as a function of information (V); one-sided SPRT test; boundaries for rejection and acceptance of null hypothesis. (From White-head, J., *The Design and Analysis of Sequential Clinical Trials*, Ellis Horwood Limited, Chichester, U.K. 1983. With permission.)

The accumulating data are plotted as a set of connected data points. If this linear representation crosses one of the boundaries as drawn in Figure 6.1, then the study is terminated and the appropriate conclusion is reached. For example, Figure 6.2 indicates a sequence of experimental outcomes which eventually leads to the rejection of the null hypothesis. At this point, the study is terminated and one concludes that the new therapy is significantly better than the standard one.

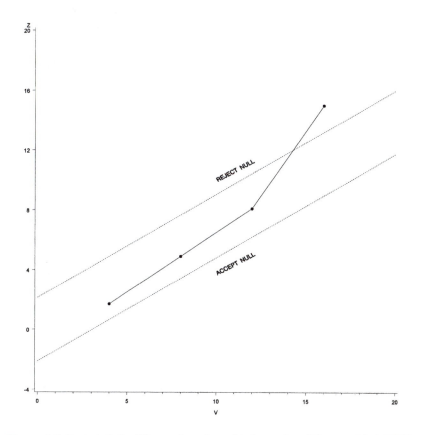

Figure 6.2 Test statistic (Z) as a function of information (V); one-sided SPRT test; data values leading to rejection of null hypothesis. (From Whitehead, J., *The Design and Analysis of Sequential Clinical Trials*, Ellis Horwood Limited, Chichester, U.K. 1983. With permission.)

As might be imagined from the form of the boundaries in Figure 6.1, i.e., parallel lines, it is possible for the test statistic and its graphical representation to wiggle around in the continuation region indefinitely. This is the open-ended property of the SPRT mentioned earlier. The probability of a very long study would be greatest when the true state of nature was midway between the null and postulated alternative hypotheses.

To overcome this deficiency, Whitehead (1983) developed a closed form sequential test called the triangular test. As can be imagined from the name of the test, the continuation region is triangular. As depicted in Figure 6.3, the triangular test has a continuation region that is wider that that for the SPRT early in the study, but narrows down and ulti- mately converges to a single point, so that the study cannot go on indefinitely. Figure 6.3 also indicates the form of the acceptance and rejection regions for a comparable fixed size test. As indicated in the

figure, the fixed sample test has a fixed sample size or constant level
of V. The rejection region is the set of all Z values above the diamond
($\Diamond$) on this line.

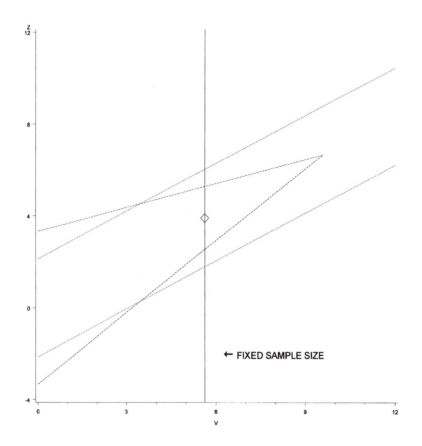

Figure 6.3 Test statistic (Z) as a function of information (V) comparison of one-
sided SPRT, triangular, and fixed sample tests. (From Whitehead, J., *The Design
and Analysis of Sequential Clinical Trials*, Ellis Horwood Limited, Chichester, U.K.
1983. With permission.)

Up to now, our discussions of sequential tests have been limited
to one-sided alternatives. If, in fact, the alternative is two-sided, then
there are more boundaries to consider and the form of the continuation
region is slightly different from those illustrated in Figures 6.1 to 6.3.
In the context of the two therapies, there are four possibilities at each
interim analysis:

1. The null hypothesis of equality of the two therapies is rejected
   and it is concluded that the new therapy is superior

2. The null hypothesis is rejected and it is concluded that the new therapy is inferior
3. The null hypothesis is accepted (and it is concluded that the two therapies are equivalent)
4. There is insufficient evidence to reach a conclusion and so the study continues

The extension of Whitehead's triangular test to the two-sided alternative hypothesis is called the double triangular test. The continuation region of the double triangular test is formed by taking the triangle for the one sided test as in Figure 6.3 and reflecting it through the horizontal axis to obtain a region as depicted in Figure 6.4. As with the previous figures illustrating sequential studies, one starts at the left side of the figure. As data accumulate in the study, one moves to the right. If a boundary is crossed, then the study may be terminated and the conclusion indicated in the figure is the result. Until a boundary is crossed, one is said to be in the continuation region.

Example 6.2 (Continuation of Example 5.4)

The veterinary study of the temporary oxygen carrying blood replacement (Section 5.2) was conducted as a sequential study using a modified double triangular test (Selwyn and Rentko, 1995) to compare the success rates between treated and control dogs. In order to accumulate sufficient safety information on the product, it was determined that the first interim analysis would be conducted after 60 dogs completed the study. Subsequent interim analyses were to be performed after every 10 additional dogs thereafter. Figure 6.4 presents the findings from the efficacy population* from the study starting with a total sample size of N = 20 and then in increments of 10 until the final sample size is achieved. Starting with N = 20 is only depicted for illustration. In actual practice, no analysis would have been performed prior to N = 60 because that was the restriction specified in the study protocol.

Superiority of the active treatment was clearly demonstrated early in the study, even after 30 dogs in the efficacy population had been tested (see Table 6.1). Additional patients only confirmed this finding. If this study had been run as a strictly sequential study with interim analyses every 10 completed patients, it would have been able to terminate at N = 30.

The reason that a sequential design was very effective in this study was that the initial estimates of success rates for the two competing treatments used to determine sample sizes for a fixed size trial were

---

* In the context of clinical trials, an efficacy population is often distinguished from an intent-to-treat population. The intent-to-treat population includes all patients randomized to treatment. The efficacy population often includes only those patients who (1) meet all inclusion and exclusion criteria, (2) do not experience any serious protocol violations, and (3) do not drop out of the study so early that their data are irrelevant.

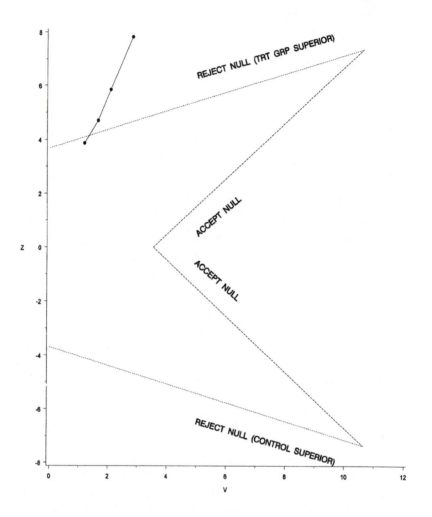

Figure 6.4 Test statistic (Z) as a function of information (V); veterinary study; efficacy population.

not accurate. The original estimates of success rates in the treated and control groups made prior to the study were 80 and 50%, respectively. Our sample size estimate for a fixed-size trial using a false positive rate of 5% for a two-sided test and a power of 80% was 45 dogs per group for a total sample size of 90. The actual success rates in the efficacy population for this study were 95 and 32%. These are considerably farther apart than originally imagined. Hence, our use of the modified sequential test was very successful.

This example also brings out one more idea that was perhaps not fully explored in Chapter 4. While sample sizes are determined on the

Table 6.1 Test statistic (Z) as a function of information (V); veterinary study efficacy population

N	Z	V
20	3.88	1.19
30	4.71*	1.65
40	5.86*	2.08
50	7.83*	2.85

Note: The test statistic Z tabled here and depicted in Figure 6.4 is not yet "standardized" and does not correspond to the standard normal distribution under the null hypothesis. To standardize, Z must be divided by the square root of V. This will result in a statistic that is standard normal under the null hypothesis of no treatment effect. Asterisks in the table identify values which lie beyond the boundary for study continuation and would typically lead to study termination. For example, it can be seen from the above table that the study could have been terminated at N = 30 with the conclusion of statistically significant (0.05 level) efficacy of the drug.

basis of the best information available prior to a particular study, such information is often based on some guesswork. It is therefore critically important to conduct sensitivity analyses on such sample sizes (Section 4.7). Sequential designs such as the one illustrated here can be extremely useful in situations in which little is known *a priori* about the underlying biological or medical phenomena and where experimental outcomes are subject to some speculation. When actual results are much better than expected, sequential trials can stop early with good results. When actual results are much worse than expected, sequential trials can stop early with good savings. When results are about as expected, designs such as those corresponding to Whitehead's triangular test rarely cost more than their fixed size counterparts. However, such designs are not always applicable and their utility depends in part on the possibility of determining each patient's primary outcome rapidly relative to patient recruitment.

6.4 Group Sequential Designs

While the advantages of a sequential design may be clear, such a design is not always possible or even practical. For one thing, the endpoint utilized in assessing early stopping must be one that is obtained relatively quickly for each experimental subject relative to the rate of accrual in the study. Second, for many investigators (and statisticians), it is not necessarily convenient to analyze data after every new subject, pair of subjects, or even at regular frequent intervals as is the case in a purist sequential analysis. This latter consideration is especially true in multicenter clinical trials in which the logistics of obtaining data from several centers simultaneously makes frequent updating and analysis of data very difficult. To overcome these practical limitations on sequential designs yet still retain the benefit of possible early stopping in clinical studies, statisticians have developed an alternative set of designs called group sequential designs.

In group sequential designs, the number of interim analyses is fixed in advance at a relatively small number, usually between two and five, with a final analysis being included in this total.* At each interim analysis, we can either reject the null hypothesis or continue to collect data. At the final analysis, either the null hypothesis will be rejected or it will be accepted.

Sample size estimation is still necessary even with group sequential designs. Once the total sample size is determined, interim analyses are conducted after a fixed number of subjects have completed the study. For example, suppose that the total sample size in the study is 200 and that a five-stage design is selected. Interim analyses would be conducted after 40, 80, 120, and 160 patients as necessary. If a decision was not reached by the fourth interim analysis, the final analysis would be conducted after all 200 patients completed the study.

Group sequential designs differ in how the overall false positive rates are allocated among the interim analyses and whether the alternative hypothesis being considered is one sided or two sided. One can understand the difference between these plans by taking into account the situation that one might be faced with at a final analysis. Suppose that the design we are using is one in which we have used up most of the false positive rate early in the study because we had hoped that the effect we were seeking would be evident early in the study. As a result of our early spending, we have to be more conservative at the end of the study and require greater evidence (i.e., a more extreme test statistic) to be able to reject the null hypothesis at the final analysis.

On the other hand, suppose that we have come to the final analysis, but have spent very little of the overall false positive rate in previous interim analyses. Now we may have almost all remaining and it will take less evidence (i.e., a test statistic that is not quite as extreme as the case above) to reject the null hypothesis. What we have sacrificed, however, is much of the chance for early stopping that we might have had with an "early spending plan."

Critical values (see Section 4.4) for group sequential designs with various degrees of early spending are presented in Tables 6.2 and 6.3. These tables are modified forms of tables previously presented by Geller and Pocock (1988). In Table 6.2, we display critical values and corresponding nominal significance levels for an overall 0.05 level test

---

* There do not seem to be any practical advantages to having more than five interim analyses.

assuming a normally distributed test statistic* and a one-sided alternative. Table 6.3 presents similar information, but for a two-sided alternative.

The first column in each table gives the number of stages in the group sequential design. This also equals the number of interim analyses, including a final analysis. Each successive column gives the critical value to be used at each interim or final analysis and the corresponding nominal significance level. Interim analyses are typically conducted at equal intervals (in terms of number of patients) throughout the study.

For example, in Table 6.2, the column labeled "Pocock" provides information on a group sequential design proposed by Pocock (1977). This design incorporates equal critical values and hence, equal spending per stage. With two stages, the critical value for each stage is 1.875 and the nominal level corresponding to this is 0.030. Although such a design will have the advantage of a high probability of early stopping, less of the false positive rate is retained for the final analysis.

In contrast, the other group sequential plans in the tables are more conservative in the early stages and hence retain most of the false positive rate for the final analysis. For example, the column labeled "Haybittle; Peto" indicates that if early analyses are conducted at the 0.001 level, then the final analysis can be conducted at nearly the nominal 0.05 level while still maintaining an overall false positive rate of 0.05 (Haybittle, 1971; Peto et al., 1976). The remaining columns in the table present group sequential designs due to O'Brien and Fleming (1979) and Fleming et al. (1984) which are intermediate between the very conservative approach by Haybittle and Peto and the more liberal approach of Pocock.

In choosing among the various group sequential plans presented here (and perhaps others), one should consider the feasibility of conducting several interim analyses and the advantages and disadvantages of each extra stage as well as whether a more liberal or more conservative plan is desired. As is often true, the biggest benefit is obtained when the design is modified from a fixed sample design (i.e., one stage) to a two-stage design. There are then diminishing returns with each additional stage.

Example 6.3

In a veterinary study of the effects of a novel nonsteroidal anti-inflammatory drug (NSAID) in dogs with degenerative joint disease,

---

* Although we have not previously defined the normal or Gaussian distribution, most readers will be familiar with it. It has many applications in statistical data analysis in particular because averages of independent variates tend toward the normal (the Central Limit Theorem in statistics). The reader may consult almost any introductory statistics text for a thorough description.

Table 6.2 Critical values and corresponding significance levels of one-sided group sequential tests at 0.05 level

Number of stages	Pocock	Haybittle; Peto et al.	O'Brien and Fleming	Fleming et al.
2	1.875	3.090	2.373	2.170
	0.030	0.001	0.009	0.015
	1.875	1.647	1.678	1.717
	0.030	0.050	0.047	0.043
3	1.992	3.090	2.961	2.326
	0.023	0.001	0.0015	0.010
	1.992	3.090	2.094	2.220
	0.023	0.001	0.018	0.013
	1.992	1.649	1.710	1.739
	0.023	0.050	0.044	0.041
4	2.067	3.090	3.466	2.475
	0.019	0.001	0.0003	0.007
	2.067	3.090	2.455	2.383
	0.019	0.001	0.007	0.009
	2.067	3.090	2.001	2.294
	0.019	0.001	0.023	0.011
	2.067	1.651	1.733	1.731
	0.019	0.049	0.042	0.042
5	2.122	3.090	3.912	2.570
	0.017	0.001	0.00005	0.005
	2.122	3.090	2.766	2.493
	0.017	0.001	0.003	0.006
	2.122	3.090	2.258	2.418
	0.017	0.001	0.012	0.008
	2.122	3.090	1.955	2.337
	0.017	0.001	0.025	0.010
	2.122	1.653	1.752	1.728
	0.017	0.049	0.040	0.042

From Geller, N. L. and Pocock, S. J. "Design and Analysis of Clinical Trials with Group Sequential Stopping Rules," in *Biopharmaceutical Statistics for Drug Development*, Marcel Dekker, New York, 1988. With permission.

Table 6.3 Critical values and corresponding significance levels of two-sided group sequential tests at 0.05 level

Number of stages	Pocock	Haybittle; Peto et al.	O'Brien and Fleming	Fleming et al.
2	2.178	3.290	2.797	2.432
	0.029	0.001	0.005	0.015
	2.178	1.962	1.977	2.036
	0.029	0.050	0.048	0.042
3	2.289	3.290	3.471	2.576
	0.022	0.001	0.0005	0.010
	2.289	3.290	2.454	2.493
	0.022	0.001	0.014	0.013
	2.289	1.964	2.004	2.059
	0.022	0.050	0.045	0.039
4	2.361	3.290	4.049	2.713
	0.018	0.001	0.00005	0.007
	2.361	3.290	2.863	2.641
	0.018	0.001	0.004	0.008
	2.361	3.290	2.338	2.567
	0.018	0.001	0.019	0.010
	2.361	1.967	2.024	2.051
	0.018	0.049	0.043	0.040
5	2.413	3.290	4.562	2.801
	0.016	0.001	0.000005	0.005
	2.413	3.290	3.226	2.743
	0.016	0.001	0.001	0.006
	2.413	3.290	2.634	2.684
	0.016	0.001	0.008	0.007
	2.413	3.290	2.281	2.617
	0.016	0.001	0.023	0.009
	2.413	1.967	2.040	2.051
	0.016	0.049	0.041	0.040

From Geller, N. L. and Pocock, S. J. "Design and Analysis of Clinical Trials with Group Sequential Stopping Rules," in *Biopharmaceutical Statistics for Drug Development*, Marcel Dekker, New York, 1988. With permission.

a two-stage group sequential design was selected. In coming to this decision, we first considered the sample size necessary in a fixed trial design. The study was to be a completely randomized design with three groups: placebo-treated dogs, dogs administered a low dose of the NSAID, and dogs administered a high dose of the NSAID. The critical comparison in the study was between the placebo and high-dose groups.

Based on the estimated magnitude of the treatment effect and its variance provided by data from a pilot study, we calculated that 40 dogs per group (120 dogs total) were required to provide 80% power based on an analysis of variance for all three groups. Instead of using a fixed-size design, however, we decided to employ a two-stage group sequential design with equal critical values per stage (Pocock, 1977) with boundaries as provided in Table 6.2. Computer simulations conducted prior to the start of the study indicated that there would be a 57% chance of finding a treatment effect (i.e., rejecting the null hypothesis) at Stage 1, and an 86% chance of finding a treatment effect by Stage 2 (the cumulative probability from Stages 1 and 2) assuming the magnitude of the effect seen in the pilot study and the same variation.*￼ As of the writing of this chapter, this study is still ongoing.

There are some other aspects of this study that are worthy of note. In this study, there were two major types of responses. Objective responses were determined from *force plate* measurements. In order to obtain the force plate measurements, dogs were led on a leash in such a way that they would step on a plate which would automatically record various aspects of measurements related to gait. During each visit (two baseline visits and a visit after one week of treatment), five replicate trials were made by walking the dog across the force plate path five times. Data from the five replicate trials were then averaged to derive a reliable estimate of performance. A single trainer was used to make all of the trials for each dog on all three visits. The reason for this was to eliminate "trainer" as a source of variation for each dog. In addition, prior to the first baseline visit, a training session was held so that each dog became acclimated for the force plate measurements. One of the force plate measurements was identified based on results from the pilot study as the primary efficacy endpoint and used as the basis of the interim analysis. The other force plate measurements were considered secondary efficacy endpoints.

Several subjective measurements were also made during this study. For example, a lameness grade was assigned to each rear limb at each

---

* The power calculations comparing the two-stage group sequential test (86%) to the fixed sample test (80%) are not quite fair. The former utilizes only the comparison between the high dose group and the placebo group. The latter is based on all three groups and therefore has lower power when the effect in the low dose group is between the other two, the assumption made in the calculation.

visit based on an ordinal grading scheme. The categories were identified as none, mild, moderate, and severe lameness. Definitions were provided in the study protocol. The individuals conducting the force plate measurements as well as the clinicians making the subjective assessments were blinded as to the treatment group status of the dogs in the study.

6.5 Interim Analyses

There is considerable misunderstanding regarding when and how interim analyses can be conducted in medical studies. If the question is "When can interim analyses be conducted?", then the answer is "As often as one likes, as long as the interim analysis results are not used to make substantive changes in the study" such as terminating the study and concluding treatment efficacy. Thus, in theory, interim analyses can be conducted very frequently as long as no action is taken based upon them.

In practice, however, it would be very difficult to conceal the results of interim analyses from eager investigators or study sponsors. Despite the best of intentions, it is likely that such information will leak and there may be considerable pressure on study planners to terminate such a study early or at least to consider this possibility. This is certainly an ethical dilemma. Suppose that one of the treatments is clearly seen to be superior. Would it not be justifiable to terminate the study at this point so that future patients could benefit as soon as possible from this new therapy and as few patients as possible would be exposed to an inferior therapy?

This is clearly a difficult issue and one that cannot be resolved on the basis of statistics alone. However, if interim analyses may be included as part of a clinical study, then it is certainly advisable to plan for them prior to the start of the study and incorporate them in the study protocol. In many situations, sequential or group sequential designs can be employed and the above ethical dilemma avoided.

6.6 Data Monitoring Boards

In clinical studies designed to incorporate interim analyses and even for studies without them, a common practice is to include an independent scientific committee to monitor the ongoing progress of the trial and to make recommendations about its conduct. Such a committee is often referred to as a data monitoring board or data and safety monitoring board. Such a data monitoring board, if established, performs a number of functions in a study (Wittes, 1993):

1. It monitors the performance of the study. This can be viewed as putting into place a set of procedures so that the study is implemented in such a way that, if the therapies being tested differ from each other with respect to safety and/or efficacy, such results will become known as a result of the trial.
2. It monitors the safety of the participants. This in itself can be a difficult task, in part because potential adverse effects of a new therapy cannot always be specified in advance.
3. It monitors for efficacy. Here, the data monitoring board may recommend terminating a study early if a particular therapy appears strongly superior. Such monitoring for efficacy is easiest for acute diseases, quickly determined endpoints, and if a program for interim analyses has been included as part of the study protocol.

As indicated in the previous sections of this chapter, statistical considerations for interim analyses and the possibility of early termination are generally straightforward, once a single efficacy endpoint and a particular design/analysis plan have been selected. However, the decision to actually terminate a study is virtually never made based on statistical considerations alone. As discussed by DeMets (1987), this would be an oversimplification of the process. In addition to the one response outcome being analyzed sequentially, there are often a number of secondary efficacy variables as well as safety variables which are being studied simultaneously. The internal consistency of results across several outcomes, both primary and secondary, as well as consistency within particular study subgroups and with other similar studies (if any) needs to be considered. The recommendation to terminate or continue is often made by a group of experts from various disciplines who collectively reflect on all the available data and then make an informed decision.

In discussing the issues related to the composition and role of data monitoring boards, Wittes (1993) mentions the following: whether all trials should have such boards, the responsibilities of such a board, its reporting relationship within the study, its size and composition, the expertise of its members, codes of conduct for members, relationship with other ongoing similar studies, and relationship to the study sponsor. While there is general agreement that the establishment and productive use of data monitoring boards in clinical studies are beneficial, there is not consensus about many of the above issues.

Fleming (1993) points out an important point about data monitoring boards, i.e., that the members should be ethically and scientifically supportive of the protocol objectives and the design of the study. Further, they should have confidence in the procedures that are planned for capturing all study data and information. Through its review of accumulating study data and information, the data monitoring board will not only be able to make recommendations about when to terminate a study early, but also be in a strong position to make recommendations about modifications in study design or conduct, data management and quality control procedures, and/or reporting methods. By establishing a data monitoring committee with responsibility of interim reviews, the overall quality of the study and its data will be substantially enhanced.

There are major differences between data monitoring boards in studies sponsored and conducted within the pharmaceutical industry and for cooperative clinical trials sponsored by government agencies like the National Institutes of Health (NIH) and the Veterans Administration (VA). Rockhold and Enas (1993) indicate that external data monitoring boards are seldom employed for studies conducted by industry as the resources to monitor such trials are often contained within the companies themselves. The data monitoring function (which Rockhold and Enas distinguish from the interim analysis function) is said to be a continuous exercise conducted by the clinical research department for every study run. However, in some instances, independent data monitoring boards are utilized. Rockhold and Enas recommend an external data monitoring board for studies in which there is the potential for severe adverse events or if mortality or a life threatening endpoint is being studied, or if the study is being contracted out by the sponsor.

Walters (1993) indicates that a data monitoring board should have the maximum feasible independence. He argues that individuals with direct involvement for treatment of patients in the study should be excluded from a board (as is always the case), but also that representatives of the sponsor and regulatory agencies should be excluded as well. He maintains that the review of clinical outcomes should be conducted in a closed session by data monitoring committees and the statistical groups supporting them. Inclusion of sponsor representatives might allow the business interests of the sponsor to influence the committee's decisions. Moreover, an independent committee would contribute to the quest for knowledge by having freedom from a variety of interests and motivations. The credibility of a commercially funded trial would also be enhanced if the sponsor appoints an external statistical center which also maintains an arms-length relationship to the sponsor.

In contrast, Enas and Zerbe (1993) propose a very different composition for data monitoring boards in studies conducted by the pharmaceutical industry. They recommend that the typical board have representation from the sponsor's senior management, study monitors, statisticians, and therapeutic experts either inside or outside the sponsoring organization. In their paradigm for interim analyses, they appropriately argue that a major facet of interim analyses is the documentation of the process and its impact on the conduct of the study. As I have indicated above, scheduled interim analyses should be planned as part of the study protocol. In addition, the interim databases should be archived so that interim analysis results can be traced back in time. Standard operating procedures should be developed and followed which document the process and its results. Such procedures help to maintain the credibility of the trial.

In summary, data monitoring boards can provide a useful mechanism to help ensure the quality of information and decisions arising from clinical studies. While there may be some disagreement about the composition and the exact role of such a committee depending on one's experience and outlook, it is generally recognized that such data monitoring boards can be extremely important, especially in large and difficult studies. This is particularly true when interim analyses of study data are conducted periodically throughout the course of the study. I believe that many such studies will benefit by the establishment and utilization of such committees and that their function and scope of involvement in decision making should be specified in the study protocol when such an approach is implemented.

chapter seven

High Dimensional Designs and Process Optimization

In this chapter, we discuss several very interesting topics in experimental design. In particular, we consider the situation where, at the early stages of a research project or investigation, one is faced with many factors under study. It would often be impractical to design a full factorial study in this situation. Instead, we introduce the notion of fractional factorial designs in which the total number of treatment combinations actually included in an experiment is realistic, yet one in which all factors are still simultaneously studied.

Suppose that one has determined which experimental factors are of most interest. We may be interested in optimizing some response variable as a function of the levels of these factors. For example, in a production process we may wish to maximize the yield or purity of our product and want to find the combination of levels of each factor which will allow us to do so. This can be accomplished using response surface methodology and/or process optimization methods as discussed in this chapter.

Thus, we view the various methods discussed in this chapter as potentially related. With fractional factorial designs or other similar strategies for narrowing down a large number of possibly important factors, we are able to focus on a small subset of important factors. With response surface methodology or other methods for process optimization, we strive to find the combination of levels of these factors providing us with the optimum response.

7.1 Fractional Factorial Design

In Chapter 5 we discussed factorial designs as a special case of the completely randomized design. Suppose that we are at an early stage of the research process and there is still considerable uncertainty as to which of many factors might affect an experimental outcome. Because the total number of treatment classifications in a factorial study is the product of all levels of all factors, a study can become prohibitively large, even if each factor is assessed at only a few levels. For example, suppose that we have eight different factors under consideration and that each is to be assessed at only two levels. The total number of treatment combinations will be $2 \times 2 \times 2 \times 2 \times 2 \times 2 \times 2 \times 2 = 2^8 = 256$. With very few exceptions, this would be a completely infeasible study to perform.

In order to fashion a design to allow us to study all these factors simultaneously, but without making the total number of treatment combinations unreasonable, we only include a systematic fraction of all the possible factorial treatments. Hence, the design is called a frac-tional factorial design.* For example, if we were to include one half of all the possible treatment combinations in a 2^6 factorial design, then the resulting fractional factorial design would be called a one half replicate of a 2^6 design. If we included only one fourth of all possible treatments (again in a systematic way to be described below), then the resulting design would be called a one fourth replicate of a 2^6 design. The former design would have a total of 32 treatment combinations (one half of 64), whereas the latter design would have a total of 16 (one fourth of 64).

As can be imagined, if we reduce the total number of treatment combinations in the study, we must pay some penalty. In fact, the penalty that we pay is that certain high level interactions (usually of degree three or higher) cannot be assessed with this type of design. They are said to be confounded or purposely mixed up with other main effects or lower order interactions. To illustrate, suppose that an observed difference in a study can either be attributed to a single factor, say A, or to some complex interaction, such as the $C \times D \times E$ interaction (an interaction between the three factors C, D, and E). It is certainly simpler and probably more realistic to consider that the difference we are seeing is due to A and not to this high level interaction.

To construct a fractional factorial design, we purposely confound main effects and low level interactions with very high level interactions. The effects that are confounded are said to be aliases of one another. For instance in the example given just above, we say that the effect of the main factor A is confounded with the $C \times D \times E$ interaction. Because these effects are completely confounded, we cannot distinguish whether a possible response is due to one or the other, but in order to

* In a situation like this, many experimenters will be tempted to conduct studies varying one factor at a time in order to determine which ones affect the experimental outcome. However, in comparison with full and fractional factorial designs, this approach has the clear disadvantage that interactions between factors cannot be assessed at all.

make the most sense out of experimental results, we attribute findings to the main effects and low level interactions first, rather than high level interactions.

As another example, in the one half replicate of a 2^6 design described below, the effects due to treatment A are completely confounded (mixed up with, indistinguishable from) the effects of the B × C × D × E × F interaction. It is certainly easier and generally more sensible to attribute the response to the single factor A than to this five-way interaction.* We therefore conclude that the factor A has an effect and ignore the possibility that the response is due to the high level interaction.

There is clearly some danger in doing this. We may be misled by the simplicity of a single factor relative to the complexity of the high level interaction. In cutting down on the total number of treatments studied in a fractional factorial design, we limit our ability to evaluate high level interaction effects. However, it is rare that interaction effects with more than three factors are meaningful, so that the use of the fractional factorial design can be very efficient when the potential number of factors in a study is large.

In order to define a fractional factorial design, we need to specify two things: (1) which specific treatment combinations will be included, and (2) as a result of these choices, which effects (i.e., main effects and their interactions) will be confounded with others. Clearly, we would not want to confound main effects with other main effects (e.g., A with B). We would probably not even be satisfied to confound a main effect with a two-factor interaction (e.g., A with B × C). However, if a main effect were confounded (an alias of) a very high level interaction, this would not trouble us so much.

To start with, we make one (or more) of the high level interactions a *defining contrast*. This allows us to determine which effects will be completely confounded (i.e., inseparable from others). The use of a defining contrast is most easily illustrated in situations where each factor is to be tested at exactly two levels. If there are K such factors, then a full factorial experiment would be from the 2^K *series* of designs. Suppose, for example, that we are considering an experiment with six separate treatment factors, A, B, C, D, E, and F. Each factor will be either present or absent and all treatments can be potentially tested simultaneously. As indicated above, a full factorial design would require 64 treatment combinations.

A one half replicate of this design would utilize 32 of the possible combinations. The best defining contrast would use the highest interaction present (the six way interaction ABCDEF). To see which effects are confounded with one another, we write down each one along with

---

* I doubt that I could explain to anyone what a five-way interaction represented anyway.

the defining contrast. We cross out all treatments (letters) that appear twice. What remains is an alias of the effect that we started with and is completely confounded with it.

For example, to see what is confounded with the main effect A, we write down the letter A and the defining contrast ABCDEF:

A   ABCDEF

The rule that we now use is to cross out any letter that appears twice. Because A appears twice, we cross it out and we are left with the BCDEF interaction:

~~A~~   ~~A~~BCDEF

Thus, the main effect of A is totally confounded (is an alias of) the BCDEF interaction in this design. Therefore, if we observe a response as a result of this treatment (it could be due to the effect of A or the effect of B, C, D, E, and F administered together), we attribute it to A rather than to the five-way interaction BCDEF. It can be seen that with ABCDEF as the defining contrast, all main effects are confounded with five-factor interactions, all two-way interactions are confounded with four-way interactions and all three-way interactions are confounded with others. To continue the example, consider the A x B x C interaction. With what is this confounded? Again, write it down along with the defining contrast:

ABC   ABCDEF.

Because each of the letters A, B, and C, appears twice, we cross them out and are left with the term DEF:

~~ABC~~   ~~ABC~~DEF.

Thus, the effects of the three-way interaction for A x B x C are completely confounded with the three-way interaction D x E x F. Either we ignore all three-factor interactions (and higher) or we give up the possibility to distinguish between pairs of them.

One way to determine which treatments and combinations to include in our experiment is to use all treatment combinations which contain exactly 6, 4, 2, or none of the letters A, B, C, D, E, and F. Thus, the treatment combinations applied in this one half replicate (one particular experiment) would be ABCDEF, ABCD, ABCE, ABCF, ... , CDEF, AB, AC, AD, AE, AF, ..., EF, and control (vehicle or placebo). The other half replicate (another design choice potentially used in a different experiment) would include all treatment combinations with an odd number of letters.

Using a one half replicate of a 2^6 design, we can assess the effects of all main effects (A, B, C, D, E, and F) and all two-way interactions (AB, AC, ... , EF). All three level and higher interactions are completely confounded and hence not estimable by themselves.

In order to illustrate this process further, consider a one fourth replicate of the 2^6 design. This might be a design of more practical interest as the number of treatment combinations is reduced from a still unwieldy 32 to a more manageable 16. In order to create the one fourth replicate, we need to identify a second defining contrast. One might think offhand that the best choice would be another very high level interaction, such as the five-way interaction ABCDE. However, this turns out to be ill-advised, as the main effect for the factor F now cannot be assessed (Cochran and Cox, 1957, p. 250). To see this, recall how we chose the treatment combinations for our one half replicate of the 2^6 design. We selected all treatment combinations with an even number of letters (0, 2, 4, or 6). This gave us a total of 32 treatments in the study.

Suppose now that we wanted to take a one half replicate of this design using ABCDE as the defining contrast. In a similar way, we could select from the 32 treatments all those which contained an even (odd) number of letters in the set ABCDE. Specifically, these would be ABCD, ABCE, ABDE, ACDE, BCDE, AB, AC, AD, AE, BC, BD, BE, CD, CE, DE, and placebo. Notice that none of these treatment combinations includes the factor F, so that its effect cannot be estimated. If we had picked those treatment combinations with an odd number of letters in the defining contrast, then all resulting treatments would include F and its main effect could still not be estimated (as every treatment combination tested would have the factor F in it). As explained by Cochran and Cox (1957), when two contrasts are used to define a one fourth replicate (in the 2^K system), their generalized interaction (i.e., the result of writing down both contrasts and striking out factors appearing twice) is also a defining contrast and cannot be estimated.

To continue with this example, we see that the two initial defining contrasts, ABCDEF and ABCDE, result in the additional defining contrast F:

$$\text{A\!\!\!/\,B\!\!\!/\,C\!\!\!/\,D\!\!\!/\,E\!\!\!/} \qquad \text{A\!\!\!/\,B\!\!\!/\,C\!\!\!/\,D\!\!\!/\,E\!\!\!/F}$$

To overcome this problem, i.e., that the main effect of factor F cannot be estimated with the one fourth replicate design described above, we instead use two four-factor interactions as defining contrasts. If we choose them in such a way so that they have two letters in common, then all main factors in the study can still be assessed. For example, we could select ABCD and ABEF. The final defining contrast would then be CDEF:

$$\overline{AB}CD \qquad \overline{AB}EF,$$

another four-way interaction. We can now construct a design with just 16 treatment combinations, but as expected, we must pay some sort of price for reducing the total number of combinations under study and hence, the total sample size. In fact, each factorial effect that is not a defining contrast now has three aliases (based on the three defining contrasts listed above). These are found in the same way as described above, i.e., by writing down each term along with all three defining contrasts one at a time. Again, any letters appearing twice are crossed out. Each term obtained in this way is an alias of the original term.

For the three contrasts ABCD, ABEF, and CDEF, the set of aliases for the one fourth replicate design is given in Table 7.1. As indicated in the table, all main effects are confounded with either three- or five-factor interactions. All two-factor interactions are confounded either with other two-factor interactions, four-factor interactions, and/or the six-factor interaction. The remaining three-factor interactions (those not already identified previously in the upper part of the table) are confounded with the others. Thus, if it is safe to ignore all three-way and higher interactions, all main effects can be uniquely assessed. However, all two-way interactions are confounded with at least one other two-way interaction. The interactions for A × B, C × D, and the E × F effects are all aliases of one another. The most prudent course of action is probably to attempt to estimate the main effects of each of the factors and ignore interactions completely in the present study. Once the set of important factors is reduced, the experimenter can run a full factorial experiment on the remaining set of factors to assess potential interaction.

One choice for the set of treatment combinations in the one fourth replicate of this 2^6 design would be: Vehicle, AB, CD, EF, ACE, ACF, ADE, ADF, BCE, BCF, BDE, BDF, ABCD, ABEF, CDEF, and ABCDEF.

This extended example points out both the benefits and the disadvantages of fractional factorial designs. We can simultaneously assess the effects of many factors in an economical way relative to the total possible number of treatment combinations, but only with the loss of information about factor interactions. The smaller the design is, the fewer the interactions we can assess. Fractional factorial designs in other series such as 3^K (e.g., a 3 × 3 × 3 design would include three factors each at three levels; one might consider a one third replicate of this design) and 4^K and even mixed series are less common. Information on such designs can be found in the book by Cochran and Cox (1957) and other sources. For example, John (1970) discusses how a fractional experiment can be obtained from a 4^3 design (assuming no interaction

Table 7.1 Confounding in a one fourth replicate of a 2^6 design with ABCD, ABEF, and CDEF as defining contrasts

Effect	Defining Contrasts		
	ABCD	ABEF	CDEF
		Aliases	
A	BCD	BEF	ACDEF
B	ACD	AEF	BCDEF
C	ABD	ABCEF	DEF
D	ABC	ABDEF	CEF
E	ABCDE	ABF	CDF
F	ABCDF	ABE	CDE
AB	CD	EF	ABCDEF
AC	BD	BCEF	ADEF
AD	BC	BDEF	ACEF
AE	BCDE	BF	ACDF
AF	BCDF	BE	ACDE
BC	AD	ACEF	BDEF
BD	AC	ADEF	BCEF
BE	ACDE	AF	BCDF
BF	ACDF	AE	BCDE
CD	AB	ABCDEF	EF
CE	ABDE	ABCF	DF
CF	ABDF	ABCE	DE
DE	ABCE	ABDF	CF
DF	ABCF	ABDE	CE
EF	ABCDEF	AB	CD
ACE	BDE	BCF	ADF
ACF	BDF	BCE	ADE

among factors) by considering each four-level factor as two pseudo-factors at two levels each. The resulting experiment would have 12 design points.

Example 7.1

A pharmaceutical industry application of fractional factorial designs is given by Kabasakalian et al. (1969) who utilized a set of three fractional factorial studies to determine which of 11 other components of a throat lozenge were interfering with the benzocaine component of the lozenge. The three studies were run as a one half replicate of a 2^3 design, a one sixteenth replicate of a 2^7 design, and a one half replicate of a 2^4 design. At the end of these studies, the investigators determined that the decrease in the benzocaine content was due to three specific components. This example is interesting in that a total of only 20 design points were used among all three studies.

In Chapter 3, we argued against confounding of experimental factors with each other or with other uncontrollable variables as this introduces unnecessary bias in an experiment. In contrast, with fractional factorial designs, effects are purposely confounded in a systematic way in order to economize on the total number of treatment combinations. By doing so, we are unable to estimate the effects of high-order interactions and can only estimate main effects and possibly some low-level interactions. However, we make the study feasible to accomplish when there are many factors under investigation.

7.2   Response Surface Methodology

The goal of response surface methodology (RSM) is to determine the levels of all treatment factors producing an optimum or near-optimum response. Suppose, for example, that our goal is to find the settings for critical variables in a chemical manufacturing process which will produce the highest yield. Another application of RSM might be to find the levels of predictor variables yielding the highest potency in a particular test system for a new set of chemical compounds. We might wish to find conditions under which certain drugs or chemicals are least toxic. In any case, the goal of the research effort is to identify the set of conditions, i.e., levels of the treatment factors, which simultaneously produce an optimum response.

A variety of designs can be used to study the process. It is common to utilize a full factorial design when there are only a few factors under study. When there are many factors, we may consider a fractional factorial design as described above, but with each included treatment effect at several levels.

We model the response as a function of the levels of all treatment factors using regression techniques. Typically, the model is a quadratic polynomial with cross-product terms combining information on levels of factors two at a time included as well. In order to be valid, the optimum must be determined to be in the interior of the experimental region.

Designs other than factorials or fractional factorials are often utilized in RSM. The earliest designs, called composite designs (Box and Wilson, 1951), are constructed by adding additional design points to a 2^K factorial design. In a central composite design, additional points are added at the center of the experimental region and along each axis. For example, with two factors ($K = 2$), the design points are as displayed in Figure 7.1. The original design points are the four corners of a square with one point in each quadrant of the figure. The points on the axes have been chosen to be equidistant with the original design points with respect to their distances from the center of the experimental region (i.e., the center of the figure). Because we are modeling the response as a quadratic function of the level of each factor, we would normally require at least three levels of each factor (see Section 5.6). Thus, full factorial studies with K = 2, 3, or 4 factors (in the 3^K series) would require 9, 27, and 81 design points, respectively. In contrast, central composite designs would require 9, 15, and 25 data points in these same situations, a considerable savings.*

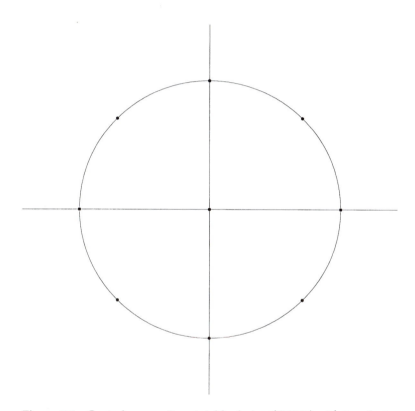

Figure 7.1   Central composite rotatable design (CCRD) with two factors.

Suppose that one has begun an investigation of the response to several factors using a 2^K factorial design. If it appears that the optimum

---

* For the two-factor case, the nine design points of the CCRD are depicted in Figure 7.1. For the three-factor case, for example, there is a total of 15 design points: 8 points which comprise the vertices of a cube centered at the origin, a center point (one additional), and 6 additional points which are the same distance from the origin as the points on the cube but lie along each factor's axis (the three factors times two directions [±] gives 6).

response is likely to be outside of the experimental region being studied, then the factorial design can be augmented with K additional design points outside of the region. One determines which design point is yielding the best response and adds additional design points along each axis from this point (see Figure 7.2 for an example with two factors, i.e., K = 2). Such a design is called a noncentral composite design.

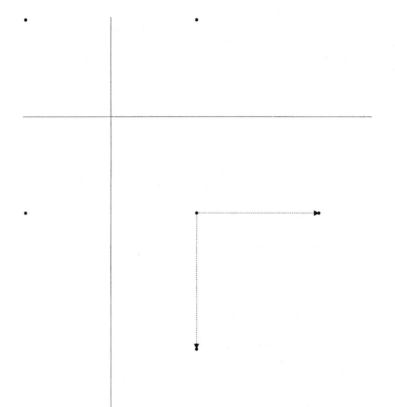

Figure 7.2 Noncentral composite design with two factors. Initial points are illustrated as the square in the upper left corner of the figure. The best response among the four points is the southeast-most point. Therefore, two additional design points are added: one in the southerly direction (lower level of factor in the vertical dimension) and one in the easterly direction (higher level of factor in the horizontal dimension).

To evaluate alternative experimental designs for RSM, one needs to consider what properties of such designs might be. One desirable property of such designs is called rotatability. A rotatable design is one in which the precision of the estimated response at any point in the experimental region depends only on its distance from the center of the region. With two factors, each rescaled so that its range is from –1 to +1, Box and Hunter (1957) have shown that a rotatable design consists of making observations at points equally spaced around the circumference of a circle and at the center itself. The points on the circumference lie at the vertices of a regular polygon (e.g., square, pentagon, etc.). In addition, the center point is generally replicated (tested more than once).

When rotatable designs are also central composite designs, they are called central composite rotatable designs (CCRDs). For two factors, such a central composite rotatable design is illustrated in Figure 7.1. Here, the center point should be replicated 5 times to give a total of 13 design points.

Example 7.2

Roush et al. (1979) employ central composite rotatable designs in the study of the protein and energy requirements of Japanese quail. In a sequence of two trials, they attempt to optimize responses (body weight and feed conversion*) as a function of the amount of protein and the calculated energy level in the diet. CCRDs as depicted in Figure 7.1 are used in both studies.

The first trial conducted was a preliminary experiment used to obtain an initial estimate of the optimum design point with body weight as the sole endpoint. Because the optimum was determined to be outside of the experimental region, a second follow-up study was conducted in which the ranges of both protein and energy levels were modified based on the results from the first study. In the second trial, an optimum was located within the experimental region studied. The authors concluded that RSM offered an efficient approach to the examination of the nutritional requirements affecting biological response in poultry.

Quintavalla and Parolari (1993) similarly use a CCRD with three factors (water activity, pH, and storage temperature) to study the growth of spores and vegetative cells isolated from bakery products. As a result of their investigation, they are able to determine a range of conditions (levels of the factors) which will result in a shelf life of 12 to 15 days. They also indicate how products need to be reformulated (i.e., how the water activity and pH need to be changed) to allow an extended shelf life of 30 days.

---

* Feed conversion is the amount of feed required to produce a unit increase in body weight. In the optimization process, one wishes to maximize the body weight gain and to minimize feed conversion.

An interesting application of RSM in the clinical setting is provided by Phillips et al. (1992). They report on a 3 × 4 factorial study of the joint effects of hydrochlorothiazide (HCTZ) and ramipril in the treatment of hypertension. Using response surface methodology, these authors obtain predicted responses (changes in supine diastolic blood pressure) as a function of treatment level by using a *segmented linear* model and *stairstep linear* model. These unusual models assume that the treatment effect is constant in some contiguous subregion of the experimental region.

7.3 Process Optimization

There are other approaches to optimization besides response surface methodology which is heavily dependent on the form of the polynomial regression model used to represent the response surface. In this section, we explore alternative experimental design approaches to optimization which do not rely on such regression methods to the same extent as RSM.

7.3.1 Sequential Simplex Design

The concept of a sequential simplex design can be illustrated quite easily geometrically. Suppose that it is possible to experiment in such a way that at each stage of a study, a new design point can be easily added. The sequential simplex strategy starts with a fixed number of design points. At each stage, the point providing the lowest yield (worst response) is identified. A new design point is selected based on the location of this particular point and the response is determined at the new point. The process is repeated until an approximate optimum is reached.

To start, we select $N + 1$ design points for a study with N factors. The starting points are provided by Anderson and McLean (1974, p. 363). In general, the design points are centered around a possible optimum and form a triangular region for $N = 2$, a pyramidal region for $N = 3$, and so on. To illustrate, consider the selection of design points as specified in Figure 7.3. Here, the levels of the factors have been conveniently centered and rescaled so that the initial design points (A, B, and C) form an equilateral triangle with its center point at the initial guess of the optimum. Suppose that the lowest yield is obtained for the uppermost design point (Point B). The next data point is then obtained by reflecting through (moving to) the opposite side of the triangle (i.e., to Point D), as depicted in Figure 7.3.

This process is continued until an optimum region is located. In particular, the next triangle would be formed by the points (A, C, and

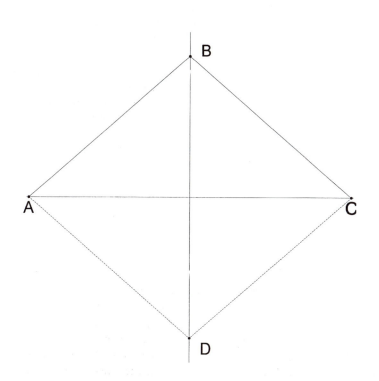

Figure 7.3 Sequential simplex design for two factors.

D). Again, we would find the point with lowest response and move away from it.

The optimum region is indicated by a collection of points which fail to produce a response that is better than one previously obtained. Anderson and McLean (1974) suggest that the presumed optimum be tested again to verify that is a true optimum and that its determination is not simply a result of experimental error.

Suppose that at a particular stage of the sequential simplex process the response at a new design point is lower that the others remaining in the current simplex. In this case, one does not go back to a previously tested point, but rather one determines the second lowest yielding design point in the current simplex and moves in the opposite direction from it.

At the end of a sequential simplex study, one may wish to perform a regression analysis using points near the presumed optimum to determine if any further improvement might be made in the yield of the process.

7.3.2   The Method of Steepest Ascent

In the method of steepest ascent, one starts with an initial design, e.g., a 2^N factorial design centered about an initial guess of the optimum. A regression model is then fit to the data from this initial design in order to determine which direction in the design space is likely to yield the most improvement in the response. One moves from the centerpoint in the design along the path of "steepest ascent." The path is then followed in steps of equal size until a decline or constant response is observed.

At this point, one may decide to start the whole process anew with another full factorial followed by model fitting and pursuit of additional points along the new path of steepest ascent. Alternatively, one may choose to employ an RSM-type design such as a CCRD and then model the results obtained at this stage using regression techniques in order to project an optimum.

7.3.3   Experiments with Mixtures

In some studies, the experimenter is faced with the situation in which the factors in a study represent the components of a mixture and the percentages of each component sum to 100%. In such cases, the allowable design points in a study are thereby constrained and the various selections of initial designs discussed above cannot be utilized. In addition, constraints on individual components may also be applicable. For example, a particular component may be limited to no more than 10% of the total.

Mixture experiments without constraints on individual components can be designed and conducted using what have been termed simplex lattice designs (Scheffé, 1958). For three components, possible designs can be depicted as points within or on the surface of an equilateral triangle. For four component mixtures, the designs are points on the surface of or within a pyramidal shaped region. As an example, Figure 7.4 depicts a seven-point design for a study with three components (factors A, B, and C). The particular design points utilized are (1) 100% A; (2) 100% B; (3) 100% C; (4) 50% A, 50% B; (5) 50% A, 50% C; (6) 50% B, 50% C; and (7) 33 1/3% A, 33 1/3% B, 33 1/3% C. The first three of these points, representing 100% for each component, respectively, are located at the vertices of the triangle. The next three points, 50/50 mixtures, are located at the midpoints of each triangle side. The last design point, representing one third for each component, is located at the center of the triangle.

Such a design can be analyzed using a regression model with each factor represented linearly and all possible cross-product terms up to the three-way included, but without any quadratic or higher terms in

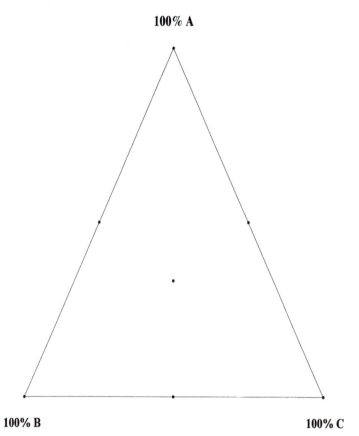

Figure 7.4   Seven point mixture design with three factors.

each factor (see Anderson and McLean, 1974). In many cases, the results may be clear-cut without any formal data analysis. Otherwise, either RSM or some form of sequential experimentation may be pursued. For a comprehensive treatment of mixture experiments, the interested reader is directed to the text by Cornell (1981).

chapter eight

The Correspondence Between Objectives, Design, and Analysis — Revisited

8.1   Data Analyses vs. Study Objectives and Design

As stated earlier in the book, there is a one-to-one-to-one correspondence between study objectives, study design, and data analysis methods. This means that the objectives will drive or determine the design, and the design and objectives will drive the analysis. Conversely, if one knows the proper analysis method to be used in a given study, it will be clear what the design is and from these, one can infer, in general, what the objectives of the study are.

This is not to say, of course, that the same statistical analysis methods can or should be applied to every data set. Another factor which determines which methods are applicable is the type of data being analyzed. This issue is discussed further in Section 8.2. Here, the main distinction is between data that are essentially continuous and data that are essentially discrete, i.e., only take on a small number of values.

Even for data of a particular type, we sometimes make assumptions about the distributional form of data and/or other relevant properties of the data. The verification of such assumptions is discussed in Section 8.3.

In the subsequent sections of this chapter, we discuss adjustments for multiple testing when many variables or hypotheses are considered in a single study, the proper use of statistical packages and cautions about these, general analysis strategies, and so-called *meta-analyses*. Meta-analyses are analyses which combine data across several related studies, even studies with somewhat different experimental designs.

Clearly, it would be impossible to cover all issues of statistical data analysis in a single chapter. It would even be difficult to do so in a single book volume. However, once an experimenter has considered the various design issues presented in earlier chapters, i.e., objective setting, development of the protocol, review of alternative designs, sample size estimation, randomization, etc., he/she may wish to have at least a fundamental idea about possible analysis methods to be applied to the resulting data. The purpose of this chapter, therefore, is to present and discuss some of the issues involved in determining a proper method of data analysis, again without mathematical details. The specifics of such methods are still important, however. Descriptions of and calculations for such methods can be found in other statistical texts (see references in Section 8.6), but are beyond the scope of the material in this book.

8.2   Types of Data

Many elementary statistical texts define four types of data: nominal, ordinal, interval, and ratio.* While this is a somewhat useful way to categorize data types, I believe that the fundamental distinction is between discrete and continuous data.** In theory, it is possible for data to take on an infinite number of values, but in practice, only a finite number of digits can be kept and hence, all data are really discrete to some extent. The fundamental issue, therefore, is the number and frequency of values which are encompassed by a set of data.

If a variable to be statistically analyzed takes on only a small number of values, say four or less, then it is clear that these data should be considered as discrete and analyzed accordingly. In contrast, if data values encompass a wide range and many different values are observed in the data set, then it is likely that such data can be considered as continuous and methods used to analyze continuous data should be applied to this variable.

Even within the categories of discrete or continuous data, different methods apply depending on the number of values taken on by discrete data and the presumed distributional form of the continuous data. For discrete data, we make this distinction between dichotomous data, i.e., data which take on only two values (e.g., success or failure; alive or dead; tumor or no tumor; etc.), and data which take on more than two

---

* Nominal data are categorized data that cannot be ordered. For example, eye color can be blue, brown, green, etc. Ordinal data are categorized data that can be ordered. For example, we can classify an adverse event or disease state as none, mild, moderate, or severe. Interval data are continuous data but for which a ratio of values is not meaningful. For example, if we think of temperature in degrees Fahrenheit, 80° is not twice as hot as 40°. Finally, ratio data are continuous data for which ratios of values are sensible, such as mass where one object can weigh twice as much as another.
** With the four-level categorization mentioned above, nominal and ordinal data would be considered discrete, whereas interval and ratio data would be considered continuous.

values. In the former case, data such as these are sometimes called binary. In the latter case, there is the distinction between ordered data (e.g., disease categories such as none, mild, moderate, or severe), referred to as ordinal data above, and data for which categories are unordered (e.g., eye color such as brown, green, or blue), referred to as nominal data above.

In considering continuous data, the main issue is whether the data are known to be from (or assumed to be from) a particular distributional form. The most commonly assumed distribution is the normal or Gaussian distribution. Most elementary statistics textbooks illustrate the normal distribution by showing the bell-shaped curve corresponding to the normal density function. In addition to the general simplicity of calculations for the normal and the great number of statistical procedures which have been developed based on normal theory (e.g., *t*-tests, analysis of variance), there is a motivation for utilizing the normal as an underlying model in many situations. If observations are independent and each comes from the same fundamental population (distribution), and each observation can itself be considered as an average of many individual components (perhaps sources of variation or error), then the distribution of each such observation will tend toward the normal as the number of components in each observation becomes large. Heuristically, the normal or Gaussian model has been found to represent well many diverse sets of data.

This is not to say that the normal distribution is the only one that should be considered when analyzing data according to a specific distributional form.* Many other distributions have been used in statistics for many years. The applicability of each in a particular situation depends on the data set at hand, what may be known about such data sets historically, and the properties of the presumed underlying distribution. In contrast, methods which do not specify a particular distribution can also be applied in many instances (see Section 8.6). Needless to say, there are advantages and disadvantages of each of these approaches (parametric, nonparametric).

8.3   Verification of Assumptions

As indicated above, many of the data analysis methods typically used in statistics depend on distributional or other assumptions about the form of the data being analyzed. Clearly, if the inference being made is sensitive to such assumptions, then the data analyst should be concerned about the validity of these assumptions. In such situations,

---

* Statistical methods which assume a particular distributional form are called *parametric* methods as the parameters of a distribution completely specify its functional form. In contrast, methods which do not depend on a particular distributional form are called *nonparametric* or *distribution-free*.

many statisticians prefer to perform preliminary tests to assess the distributional properties of data (e.g., whether the data come from an underlying normal population) or other aspects (e.g., whether all groups in a study have equal within-group variances) or to perform such tests in conjunction with their routine data analyses.

Such additional tests may or may not be helpful depending on the context. If the data set being examined is very, very large, then such tests will have extremely high power and may reject a null hypothesis even if the deviation from it is negligibly small. Conversely, if the data set being examined is small, then departures from the traditional assumptions may not be detected at all. The real issue is not only whether the assumptions being made actually hold, but how deviations from such assumptions, if present, would affect the inferences being drawn from the experimental data.

As an alternative to preliminary or post-analysis hypothesis testing of assumptions, I prefer to explore the properties of a data set through more informal methods such as diagnostic plotting. I find such plots to be useful adjuncts throughout the data analysis process. This view is also expressed by other statisticians (e.g., see Box, 1980). For example, Selwyn and Gaccione (1993) show how normal plots (plot of ordered observations or residuals) can be used to assess both normality and equal within-group variances, when several small samples arising from different populations are considered jointly. A comprehensive discussion of graphical and other techniques in the context of assessing distributional assumptions is provided by D'Agostino and Stevens (1986).

We are most concerned about verifying assumptions when it appears that the inferences from a particular study will be sensitive to them. A completely different approach to data analysis is to utilize statistical methods which are not sensitive to underlying assumptions or even moderate changes in a few data values. Such methods are called robust, as the inference being drawn is robust against departures from underlying assumptions.

Several simple examples may help to illustrate the concept of robustness. Suppose that a data set contains the three values 1, 2, and 3. Then the arithmetic mean of the data set is 2. The median of the data set (in this case the middle value after ordering from smallest to largest) is also 2. However, if we replaced the value 3 with the value 999, the mean would become 334. However, the median would still be 2. In this case, we see that the mean is severely affected (changed by a single very large or very small observation), but that the median is not. The median is in some sense, a robust estimator of the middle of the population.

Again consider the three values 1, 2, and 999. If we were to rank them from smallest to largest, their ranks would be 1, 2, and 3. Thus, it does not matter how large the largest observation in the data set is, it will always get a rank of 3 (assuming that there are no ties in the data). Thus it is clear that statistical methods utilizing the ranks of data rather than the actual values themselves (i.e., nonparametric methods based on ranks) will be robust.

8.4 Multiplicity Adjustments

In Chapter 2, we discussed study objectives and the idea of limiting the number of objectives in any study. A similar issue arises when a vast number of variables are measured in a study. Except in a study of the most exploratory nature, it would be unreasonable to statistically analyze every possible variable (including those measured at multiple time points) and be able to draw a valid conclusion on the basis of the significance of only one or a few variables. The reason for this is that the overall false positive rate across all variables and tests would be much higher than the false positive rate (alpha level) for each one. Consequently, it would be very likely that a few significant results would be obtained by chance alone. What are the approaches to deal with this multiplicity problem?

The first step is to recognize that a potential problem exists and to deal with it at the protocol writing stage. Suppose that we are developing a clinical trial protocol and want to examine the efficacy (and safety) of a new drug or treatment. We need to establish which limited set of variables will be primary efficacy variables and which (if any) will be secondary variables. If the choice among these is unclear, then perhaps we are not really at the point where we can conduct a comprehensive study. Instead, we may wish to conduct a pilot study first, at least to get a preliminary indication as to which of several equally logical variables should be considered the ones of primary interest. Thus, the best approach is to determine the most important variables and comparisons beforehand, i.e., before the data are collected, and to limit the number of endpoints to be critically examined.

Despite this, there will be times when more than a very few comparisons will be of interest, and the enlightened data analyst will need to deal with the issue of a potentially elevated false positive rate. From a statistical standpoint, there are a number of ways to deal with this issue at the analysis stage. Among them are

1. Performing a global test for differences and only conducting follow-up tests if the global test is significant
2. Employing a statistical procedure which takes into account the number of hypotheses to be tested and adjusts for it accordingly

3. Testing hypotheses in a sequential manner, so that certain tests are only conducted if all previous ones are significant.*

Some examples of these three approaches would be, respectively

1. Analysis of variance (ANOVA) as a global test of group differences in the one-way layout, followed by pairwise *t*-tests if the overall ANOVA test indicates significance (this procedure is sometimes called Fisher's protected least significant difference test)
2. Multiple comparisons tests such as Tukey's, Scheffé's, Dunnett's, or Dunn's tests, or using a Bonferroni adjustment for the number of tests conducted
3. Sequential application of trend tests in a dose-response study (see Selwyn, 1995)

In summary, when a study includes a large number of variables or large number of time points, one needs to exercise care regarding the total number of statistical tests conducted. Either one needs to prioritize variables so that some small set is considered primary and all others are secondary or approaches for dealing with the multiplicity problem should be indicated in the protocol. These caveats are in keeping with the general idea of defining statistical methods before the analysis stage and including them as part of the protocol (Chapter 2).

8.5 Statistical Packages

Perhaps the biggest change in statistics in the last 30 years has come as a result of improvements in the computing hardware and software available. Statistical methods which were complex and extremely tedious to implement in the early days of the modern computer era are now easily accomplished. Problems which took minutes or even hours on early computers are now being done in seconds on a desktop personal computer (PC). Similarly, there is a wealth of statistical software available to scientists and engineers for analyzing their data.

I recall an experience in this regard in my early years in the pharmaceutical industry. A certain type of drug screening experiment was conducted in the laboratory. Typically, the experiment was run on Monday and again on Thursday. When I inquired why the experiment was not done on a daily basis, I found out that the lab technicians would spend about 2 days compiling the data, calculating summary

---

* A fourth approach might be to perform a multivariate test on a number of variables simultaneously. While such a test would suffice in terms of controlling the false positive rate, it might be difficult to interpret results, especially if variables of different kinds were being combined.

statistics, and graphing the experimental data. Data collected on Monday afternoon were processed on Tuesday and Wednesday. Data collected on Thursday morning were processed on Thursday afternoon and Friday. This left some time for experimental set-ups on Monday morning and Wednesday afternoon. All calculations were done on hand-held calculators and all graphs were manually drawn on graph paper.

With the availability and general user-friendliness of statistical packages and related hardware, such inefficiencies have presumably been eliminated forever. However, I believe that there are some drawbacks to the widespread use of statistical packages by individuals without specific training in statistics and/or data analysis. First, even as a very experienced user of statistical packages, I am aware that such software can easily be used incorrectly. Sometimes the documentation is unclear, sometimes it is wrong, and sometimes the program does not work properly (i.e., gives erroneous results, a program bug). If the user is statistically naive, then it becomes even more difficult and perhaps impossible to detect such problems.

Second, untrained users can easily perform inappropriate analyses simply because the packages allow them to do so, usually without any warnings. Earlier in the chapter I alluded to different types of data: discrete and continuous. It would be easy for a naive data analyst to apply methods suitable to continuous data only to a set of data taking on only a very few values. Except under unusual circumstances, it is unlikely that a statistical package would alert him/her to the potential dangers of doing this.

While I am certainly appreciative of today's statistical software, both as a user and as one who has been intimately involved in statistical computing for many years, I would caution the untrained individual to be aware of some of the pitfalls in their use (and abuse) and suggest several ways in which such potential problems can be overcome. First, if studies are to be replicated over a period of time and/or many similar studies of a particular kind (design) are to be performed, then it is probably worth consulting with a statistician about data analysis methods for these studies. Second, the user should carefully read through a package's documentation to assess its utility for the experimental data at hand and to become familiar with any restrictions, limitations, or constraints indicated in the documentation. Third, if a particular package is to be used routinely, then the user community within a particular organization might want to schedule a training session (sessions) on the use of that package. Fourth, in order to gain some familiarity with data input and output for a particular package, the user might run some sample problems. In particular, if the sample problems are those found in statistical texts illustrating specific methods, the user can then check program output with published results to make sure that they match, or at least to understand why they might differ. Finally, if unsuspected or unusual results are obtained with any package, this is a signal that either (1) the user has made a mistake in his/her particular problem, (2) there is an error in the computer program or its documentation, and/or (3) the data set being analyzed is unusual in some way.

Some statistical journals and computing magazines provide reviews of statistical packages. I would encourage scientists and engineers who anticipate performing their own data analyses to thoroughly explore these resources before committing to a particular package.

8.6   Analysis Strategies

In this section, we discuss statistical methods for data of various types. As explained earlier, a major distinction can be made between parametric and nonparametric methods. With regard to the parametric methods described here, the form of the underlying distribution is usually the normal or Gaussian distribution. This is not to say that other distributions are not utilized in statistics, rather that the normal is probably the model which is used most often. As indicated earlier, there are some theoretical justifications as to why the normal distribution might apply in a wide variety of situations. Even if the normality assumption does not hold, it is possible to transform data to normality by applying a mathematical transformation.

For example, if data are positive and right-skewed, i.e., have a few very large values compared to the bulk of the observations, then transformation by taking square roots or logarithms often results in a distribution much closer to the normal. A general family of power transformations has been proposed by Box and Cox (1964).

Even though our discussion of parametric methods is limited to the normal distribution, there are many other distributions used in statistical data analyses. For example, in the analysis of survival data, other common distributional forms are the exponential, the Weibull, and the lognormal, as well as several others (e.g., see Lawless, 1982).

An alternative strategy to parametric methods is provided by nonparametric or distribution-free methods. In the context to be discussed here, distribution-free methods for more than one population (sample) assume that all data come from the same unknown underlying distribution, but that the populations differ only in location and not in "shape."

The final set of methods to be discussed apply to discrete and/or categorical data. In particular, such methods are applicable to cross-classified data which are sometimes referred to as contingency table data. For example, if one outcome variable in a study is a classification of disease state as none, mild, moderate, or severe, and a primary focus of the analysis of this variable is whether two experimental therapies, say treated and control, give rise to the same frequency distributions (proportions in each state) of the disease state, then such data can be laid out in a 2 × 4 contingency table of counts as shown in Table 8.1.

Table 8.1 Example of 2 × 4 contingency table

	None	Mild	Moderate	Severe	Total
Control	*				
Treated					
Total					

Note: * Within each cell is a count of the total number of individuals in each group with that disease state.

Lastly, I mention that the discussion of statistical methods presented below is in the context of hypothesis testing. Recall that in Chapter 4 we spoke about three different statistical contexts with regard to sample size determination: estimation, confidence intervals, and hypothesis testing. Needless to say, there is a close connection between these contexts. For example, if one is performing an analysis of variance (one of the statistical techniques listed below under parametric methods), then it is natural to estimate the effect of any particular treatment category using differences of cell means, as the means are used in construction of the analysis of variance (ANOVA) table. There is also a close connection between hypothesis testing and confidence intervals in the parametric framework. The set of parameter values which would not be rejected by a hypothesis test at a certain significance level correspond to an appropriately defined confidence interval (single parameter) or confidence region (several parameters). Scientists who have constructed confidence intervals are probably more familiar with the idea of pivoting* with regard to the sampling distribution of a statistic. The duality between hypothesis testing and confidence intervals/regions provides another approach to confidence interval construction.

Even in the nonparametric framework, we can consider estimators and confidence intervals. Hollander and Wolfe (1973) provide an extensive discussion of estimators and confidence intervals related to many standard nonparametric statistical tests.

8.6.1 Parametric Methods for Continuous Data

We will assume throughout that our data are normally or approximately normally distributed, either initially or after a suitable transformation. We distinguish between several common cases: the one-sample

* The basic idea in pivoting is that working with the sampling distribution of a statistic which depends on a population parameter of interest allows one to rearrange terms mathematically to generate probability statements (i.e., confidence intervals) about the parameter.

location problem, the two-sample problem, and the more general case of a designed study with three or more treatment conditions. The methods to be discussed are outlined in Table 8.2.

Table 8.2 Parametric methods for continuous data
(hypothesis testing methods based on normal theory)

Application	Data description	Statistical method(s)	Comments
Test hypothesis about the mean of a single population	Data from a single sample	*t*-test	—
Test hypothesis about the difference in means	Paired data	*t*-test with paired data	One-sample *t*-test after taking differences for each pair
Test hypothesis about the difference in two means	Two independent samples	Two-sample *t*-test	—
Test global hypothesis of differences among means	Two or more independent samples	Analysis of variance	ANOVA provides a global test, but may not be optimal
Comparisons with a control or standard	More than two independent samples	Dunnett's test	—
Pairwise comparisons	More than two independent samples	Multiple comparison methods (see text)	—
Testing for dose-response	Several independent samples at different dose levels	Trend tests	—
Testing for treatment differences	Multifactorial or other extended design	ANOVA with additional comparisons	—

If we are dealing with a single population (sample) and wish to test the null hypothesis that the population mean is zero or some other specified quantity, then the most common normal theory test is the one-sample *t*-test. If we have paired data, so that all observations can be naturally paired, then the same approach can be applied after taking differences between the paired values. For example, a common situation in biological research is to have a pre-treatment value and a post-treatment value for each subject in the study. We take the difference between the post-treatment value and pre-treatment value for all subjects and then perform a *t*-test on the resulting difference data. This test is often called a *paired t-test*.

If we have exactly two populations and want to test the null hypothesis that the means of the two populations are the same, then this can be accomplished by a two-sample *t*-test. This assumes that all data are normally distributed and that the two populations have the same variance. One consideration with regard to the test is whether it should be one-sided or two-sided. If the alternative hypothesis is a difference in either direction, i.e., the mean for population one might either be larger or smaller than the mean for population two, then a two-sided test should be performed. If the direction for the alternative is known in advance, then the test should be one-sided. If the two populations have different variances, then some adjustment must be made to the *t*-test to account for the unequal variances and an adjusted *t*-test can be performed on these data (Welch, 1947; Fisher and Yates, 1963).

When there are two or more treatment categories or conditions, a more general formulation of a normal theory test is provided by analysis of variance. In the ANOVA framework, an overall null hypothesis is that all treatment means are the same, whereas in the alternative hypothesis we postulate that at least some means differ. If we are dealing with a multifactor experiment such as a full factorial design, then the overall difference among treatment conditions is partitioned among the various factors and their interactions (called sources of variation). A full discussion of analysis of variance for complicated designs, however, is beyond the scope of this book.

Analysis of variance is a global method for eliciting differences among treatments. Depending on the specific treatments in the study at hand and their correspondence to one another, other methods of analysis might be preferred even when normal theory methods apply. Several examples will help to illustrate this idea.

Suppose, for example, that we have conducted a study in which one of the treatments is a control or standard, to which each of the other treatment conditions is to be compared. In this case, methods of multiple comparison with a control or standard (Dunnett, 1955; Dunnett, 1964) will be preferable to ANOVA as they have been designed specifically for this purpose. Similarly, if the goal of the study is to focus on specific comparisons, such as all pairwise comparisons in the study or even just a selected set of them, then other methods of multiple comparison will generally be superior to ANOVA (e.g., Tukey's honest significant difference test [Tukey, 1953] and the Bonferroni method of adjustment for multiple comparisons [see Miller, 1981]).

Another common design is one in which increasing dose levels of a test compound are administered to different subjects, often including a zero-dose or vehicle control. If the response to the compound is likely to be more severe with increasing dose, then tests for trend are more appropriate than an overall ANOVA (see Selwyn, 1995). Such trend tests are often applied in the context of animal safety studies for new drug entities (i.e., toxicology studies).

We may also consider dose-response studies in which the goal is not only to test for a treatment (dose) effect, but also to characterize the functional form of the dose-response. In such cases, regression analysis models and methods (e.g., see Draper and Smith, 1981) can be applied to these data. Unlike ANOVA, regression methods apply when explanatory variables (sometimes called independent variables) are continuous, rather than discrete.

A final method which should be mentioned is applicable in the general context of analysis of variance, but applies when there are additional explanatory variables, often continuous, which may have an effect on the experimental outcome. In theory, such variables should not be affected by the specific treatment conditions in the study. For example, in a study with a single baseline measurement and then multiple measurements during a treatment period, we may wish to adjust our analysis on the basis of any differences between experimental units present at baseline. This is accomplished by considering the baseline measurement as a covariate and performing an analysis of covariance (ANCOVA) on the resulting data. Such covariance adjustments can be important for two reasons. First, they help to eliminate or minimize any potential bias caused by differences in the covariates, which themselves are assumed unrelated to treatment. Second, ANCOVA methods often provide increased power relative to unadjusted methods as they reduce the unexplained variation in the study.

8.6.2   Nonparametric Methods for Continuous Data

We will assume throughout that our data all come from the same underlying continuous distribution, although the distribution is not necessarily normal. We distinguish between several common cases: the one-sample location problem, the two-sample problem, and the more general case of a designed study with three or more treatment conditions. The methods to be discussed are outlined in Table 8.3.

If we are dealing with a single population (sample) and wish to test the null hypothesis that the population mean is zero or some other specified quantity, then a very common nonparametric test is the Wilcoxon signed rank test (e.g., see Hollander and Wolfe, 1973). If we have

Table 8.3 Nonparametric methods for continuous data
(hypothesis testing methods)

Application	Data description	Statistical method(s)	Comments
Test hypothesis about the location (mean) of a single population	Data from a single sample	Wilcoxon signed rank test	—
Test hypothesis about the difference in means	Paired data	Wilcoxon signed rank test with paired data	One-sample test after taking differences for each pair
Test hypothesis about the difference in two means	Two independent samples	Wilcoxon test	—
Test global hypothesis of differences among means	Two or more independent samples	Kruskal-Wallis test	Provides a global test, but may not be optimal
Comparisons with a control or standard	More than two independent samples	Dunn's test	—
Pairwise comparisons	More than two independent samples	Multiple comparison methods (see text)	—
Testing for dose-response	Several independent samples at different dose levels	Jonckheere trend test	—
Testing for treatment differences	Multifactorial or other extended design	ANOVA on ranks with additional comparisons as necessary	—

paired data, then the same approach can be applied after taking differences between the paired values. As we did for the paired *t*-test, for example, we take the difference between a post-treatment value and a pre-treatment value for each subject. We then perform a Wilcoxon signed rank test on the resulting difference data.

If we have exactly two populations and want to test the null hypothesis that the means of the two populations are the same, then this can be accomplished by a two-sample Wilcoxon test, either a one-sided or a two-sided test depending on the alternative hypothesis of interest.

A more general formulation of a nonparametric test when there are two or more treatment categories or conditions is provided by the Kruskal-Wallis test. Like analysis of variance, the Kruskal-Wallis test is a global method for eliciting differences among treatments. Depending on the specific treatments in the study at hand and their correspondence to one another, other nonparametric methods of analysis might be preferred.

In the context in which one of the treatments is a control or standard to which all others are to be compared, the nonparametric analog of Dunnett's test was developed by Dunn (1964) and is used similarly. If the goal of the study is to focus on specific comparisons, such as all pairwise comparisons in the study or even just a selected set, then other methods of multiple comparison will generally be superior to the Kruskal-Wallis test (see Hollander and Wolfe, 1973).

In dose-response studies, tests for trend should still be applied even if the normality assumption does not hold. A common nonparametric trend test is Jonckheere's test (Jonckheere, 1954; Terpstra, 1952) which can be applied in such situations.

Finally, for more complicated designs such as a full factorial design, a common approach is to rank all data for a particular variable and then to perform an analysis of variance on the ranks (Quade, 1966; Conover and Inman, 1981). In this way, we take advantage of ANOVA's ability to partition the total variation observed in a study into meaningful components (e.g., main effects, interactions), while still retaining the robust nature of nonparametric procedures based on ranks.

In choosing between parametric and nonparametric analysis methods, there are a number of factors to consider. One consideration is sample size. It is well known that, in large samples, nonparametric methods are nearly as efficient as normal theory methods, even when data are normally distributed. With very small samples, nonparametric methods have the advantage that exact critical values are often available from published tables. However, due to the discrete nature of the distribution of certain test statistics,* these small sample tests will often be conservative (i.e., operate below their nominal significance level) and hence be less powerful than their normal theory counterparts. With small to moderate samples and with an underlying normal distribu-

---

* For example, it is clear that statistics such as sums of ranks can only take on certain discrete values. As a simple illustration, consider two samples each of size two. The four observations are ranked from one to four and the sum of ranks in the first sample is calculated. The smallest value possible is 3 (1 + 2) and the largest value possible is 7 (3 + 4). This simple example illustrates the discrete nature of many nonparametric test statistics.

tion, it is also likely that normal theory methods will be superior to nonparametric methods.

8.6.3   Methods for Discrete and Categorical Data

When we speak of discrete data, we usually mean data that take on only a small number of values. Often such data can be arranged in a table of possible values, along with frequency counts for each possibility. Table 8.1 provides an example for an ordinal outcome of disease state.

Hypothesis testing with discrete data will be particular to the form of the data (see Table 8.4). For example, if the data come from a single population, then a possible null hypothesis would specify a postulated distribution of frequencies. One common null hypothesis would be that all outcomes are equally likely. Others might be probabilities based on a theoretical model. In considering the frequency distribution of the genes for eye color for children from parents, each with one brown (dominant) and one blue (recessive) gene, the three possible genotype outcomes for the offspring are (brown, brown), (brown, blue), and (blue, blue) with probabilities one fourth, one half (brown phenotype), and one fourth (blue phenotype), respectively. By observing a sample from such a population, one could test the null hypothesis of these probabilities.

In considering data from two or more populations, or in situations where data are categorized by more than one factor, the data are said to comprise a *contingency table*. The simplest such table would therefore be a 2 × 2 table. For example, one classification variable might be treatment condition (say control or treatment) and the other might be experimental outcome dichotomized into success or failure. The resulting counts would constitute a 2 × 2 table. Such a table could be analyzed by a standard chi-square test (Fleiss, 1981). Alternatively, if all row and column totals (marginals) are considered fixed, then such data can be analyzed by Fisher's exact test (e.g., see Kendall and Stuart, 1967, pp. 551–553).

In animal carcinogenicity studies to assess the safety of experimental compounds, a primary outcome is the rate of tumors in treated animals vs. the rate in control animals. In a study with just two treatments, we can analyze the resulting counts by Fisher's exact test if we regard the total number of animals found with tumor (across both groups) as fixed.

Similarly, if there are more than two treatments (say K) and the outcome of interest can be dichotomized, the resulting data can be laid out in a 2 × K contingency table. If there is no particular relation among treatments, then these data can be analyzed by a standard chi-square test to assess whether outcomes are related to treatment. If the different

Table 8.4   Methods for discrete and categorical data
(hypothesis testing methods)

Application	Data description	Statistical method(s)	Comments
Test hypothesis about frequencies in a single population	Frequency distribution data from a single sample	Chi-square test; test based on multivariate normal	Multivariate normal test for large samples only
Test for independence in a 2 × 2 table	Frequency data from a 2 × 2 table without all marginal totals fixed	Chi-square test	—
Test for independence in a 2 × 2 table	Frequency data from a 2 × 2 table with all marginal totals fixed	Fisher's exact test	—
Test for trends in proportion data	Independent proportions in dose-response study	Cochran-Armitage trend test; exact permutation trend test	—
Test for independence in an R × C table	R × C contingency table of observed frequencies	Chi-square test; exact permutation test	—
Test for lack of association between rows and columns in one or more R × C tables	One or more R × C contingency tables of observed frequencies	Cochran-Mantel-Haenszel test	—
Testing general hypotheses in multiway contingency tables	Multiway contingency tables of observed frequencies	Loglinear models (see text)	—

treatments represent increasing doses of a test compound, then such data can be statistically analyzed by a Cochran-Armitage trend test (Cochran, 1954; Armitage, 1955) or an exact permutation trend test, especially if one of the row marginal totals is small. The permutation trend test is an extension of Fisher's exact test to more than two groups.

Larger contingency tables, say with R rows and C columns, called R × C tables can also be analyzed by standard chi-square tests. The specific test will depend, as always, on both the null and alternative hypothesis involved. Several distinct possibilities may be of interest, e.g., lack of association, symmetry, etc. The specifics of any test procedure will also depend on whether one and or the other of the row and column factors is ordered or unordered. A commonly applied testing procedure for lack of association when both rows and columns are ordered is the Cochran-Mantel-Haenszel test (Mantel and Haenszel, 1959). This test applies not only to a single R × C table, but also when the results from several such tables are to be amalgamated. For example, such a test can be applied to the analysis of a dichotomous outcome (success/failure), when a number of different treatments are being compared in a single study, and when study subjects have been stratified according to another potentially important factor such as initial prognosis or disease state.

There is also a great body of literature on methods for analyzing multiway contingency tables. Again, the specific methods applicable in a given situation will depend on a number of factors such as the particular hypotheses of interest and whether categories are ordered or unordered. Extensive discussions of contingency table analyses are provided by Bishop, Fienberg, and Holland (1975) and Agresti (1990).

A final type of analysis method which has not been covered above is applicable when the response variable is dichotomous (or ordered) and explanatory variables are either discrete or continuous. Such data can often be modeled by logistic regression techniques and the dependence of outcome on the explanatory factors tested accordingly (e.g., see Hosmer and Lemeshow, 1989).

8.7 Meta Analyses

Very often, the results of several experimental studies must be considered together in some way in order to reach a sound scientific conclusion. Sometimes, this is done in an informal way by evaluating the information provided by each study and trying to assess what the total body of evidence means, much as one would take pieces of a jigsaw puzzle and see how they fit together to give a clear picture of the puzzle as a whole.

There are other approaches to combining information from several related studies, however. In particular, suppose that we are interested in making an inference about the efficacy of a drug product which has been tested in a number of studies. The easiest situation, of course, is when all the studies have been performed with a common design (protocol) with the goal of ultimately combining the data from them into an overall pooled analysis. Then, we might be able to regard each study as one which is representative of all possible similar studies. This is much akin to the concept of the randomized block design discussed in Chapter 5. Differences among studies would be considered a random effect and our analysis would center on the repeatability of results from study to study.

More often, however, we will be faced with studies with different protocols and designs, perhaps different comparisons (e.g., a placebo controlled study, an active controlled study, etc.), and different types of measurements. When the goal of an overall analysis is to quantitatively combine information from various related sources, it is termed a *meta-analysis*.

Perhaps one of the earliest methods for combining information from several independent studies (tests) is the simple method of combining independent p-values into an overall p-value for a given hypothesis (Fisher, 1932). The arsenal of statistical techniques for meta-analysis has grown considerably since such early work on combining data. In particular, sophisticated approaches based on hierarchical models (National Research Council, 1992) considering multiple levels of variation (e.g., between experiments, within experiments) and prior exchangeability can now be used to provide formal statistical analyses of experimental data from a variety of studies. For example, such a model is used (National Research Council, 1992) to show how the results from six separate clinical trials comparing the effects of aspirin to placebo on mortality rates in patients following a heart attack can be combined in a single meta-analysis. A comprehensive discussion of issues in and methods for meta-analysis is also presented in the book by Hedges and Olkin (1985).

chapter nine

Summary and
Concluding Remarks

9.1   The Role of the Statistician

Up to now, we have been describing the process of planning an experiment. In the previous chapter, we had some discussion of analysis methods as well. I would like to briefly discuss the role of the statistician throughout this process.

In the Introduction, I urged experimenters to carefully plan their experiments. I believe that this planning process can be considerably strengthened when the various statistical considerations discussed in this book are brought to bear in almost every situation. This does not mean that the planning process need be slowed down or unnecessarily burdened with restrictions, roadblocks, or bureaucratic red tape. Instead, I perceive that by carefully and routinely following the steps outlined here, the process of planning new studies can be standardized and expedited.

Statisticians can play a singular role in biomedical experimental design. Experienced statisticians will have a scope and breadth of knowledge from their practical consulting activities that will allow them to readily understand problems, suggest solutions and/or alternatives, and work closely with their biomedical colleagues to obtain the best pragmatic designs.

Recently, a potential client asked me about my experience regarding clinical studies for anti-infective therapies. I confessed that I could not recall whether I had been involved in any, but that my staff and I had considerable experience and success in assisting investigators in designing clinical studies in a wide variety of therapeutic areas. Apparently, such broad-based experience was not sufficient for this particular client, as my company was not selected for this project. What he did not realize, however, is that almost all the issues that one would face for this particular therapeutic area are generic to clinical trials in general and with such a broad base of experience, any seasoned statistician could quickly come up to speed in a particular area. Of course, statisticians with the particular expertise of interest would be at a slight advantage initially.

Consider the steps involved in the planning, conduct, analysis, and reporting of a biomedical study. These are outlined in Table 9.1. The steps in which statisticians can contribute are indicated with an asterisk. Notice that the statistician can play a vital role by providing information, advice, and assistance throughout the process. To me, it is critical that statistical input occur early in the process as such involvement will have most impact when both the objective setting and the study design are considered from a statistical point of view. It is often too late if the statistician is called in only at the data analysis phase. The study may already be compromised.

Table 9.1 Steps in the planning, conduct, analysis, and reporting of experiments

1.	Define management objectives*
2.	Define study objectives*
3.	Design experiment*
4.	Estimate sample sizes*
5.	Randomize (assign subjects)*
6.	Conduct study and collect data
7.	Perform statistical analyses*
8.	Interpret results and write reports*
9.	Present findings to regulatory agency*

* Steps in which statisticians may be helpful.

From Selwyn, M. R. *Clinical Research and Regulatory Affairs*, 9(3), 211, 1992. With permission from Marcel Dekker, Inc.

Statisticians are not experts in every aspect of medicine or biology. Here we must rely on our biomedical colleagues to inform us of the relevant medical issues and to describe to us (in laymen's terms, of course) what we are trying to accomplish in our research effort. Similarly, statisticians need to explain to their colleagues about the relevant statistical issues and decisions to be made. For everyone this is a difficult process, as each discipline has its own terminology (jargon) and methods. Despite the difficulty, this is usually a very worthwhile and rewarding experience. There is clearly a synergy (positive interaction) which takes place in the group process that brings forth the best ideas from each individual. The result is often far superior to what would be achievable without such dialogue.

Is it necessary to involve a statistician in the design of every experiment? No. Suppose for example, that a researcher is just bringing a new screening study "on-line" and has worked closely with a statistician to investigate sources of bias and variability, optimize the design, standardize the analysis, etc. It should not be necessary that the statistician review each replicate experiment as the general form of the design has been thoroughly investigated and made as efficient as possible. If the researcher decides at some future time that certain conditions of the experiment should be changed and wants help in investigating the effects of such changes, then it would seem desirable to have the help of the statistician again for that investigation. Similarly, if the experiments are expected to go on for a long time, then it might be advisable to put into place some quality control assessment mechanisms to ensure that the studies are being performed with the same high standards as they were initially. However, for standard day-to-day operation, it would probably be a waste of everyone's time to have routine participation by the statistician.

9.2   Summary

I have tried to cover a wide range of material on experimental design in this book. I am sure that some readers may find some of the issues presented here daunting and yet, some ideas will be considered obvious or even trivial to very experienced researchers. Let me try to summarize a few key ideas. If there are no other messages that the reader retains except for these, I will still consider my effort to be a partial success. The key ideas are

- Keep experiments simple

    Keep all experiments as simple as possible. Limit the number of study objectives to a manageable number. This means one, two, or at most, three. If there is a choice between a simple design and a more complicated design, choose the simple design unless there are good reasons to do otherwise.
- Develop a written protocol

    Nearly every study should have a written protocol. Studies which are designed "on the fly" are rarely successful. Even if the results from such studies appear to be promising, it is unclear that they can be duplicated because of the circuitous nature of the study design. Having a written protocol encourages all participants to perform study tasks in a uniform way. The protocol is a document that can be presented to study reviewers (e.g., regulatory agency scientists) at the end of the study to describe the experiment and document its procedures.

    Even the statistical methods used for sample size estimation, randomization, and data analysis should be specified in the protocol. These should be included to a level of detail which will allow independent experts to understand what was done and, if necessary, be able to replicate the analysis results from the study data.

- Consider all possible experimental outcomes and their interpretations

I find that a very common experience in consulting with researchers on experimental design is that they often have a preconceived notion of how a study will come out. This bias may be so strong that they refuse to accept that any other result is possible. This attitude is not only short-sighted, it is counterproductive. By refusing to accept any possible outcome except the one that they anticipate, these researchers make it extremely difficult to design good studies.

By considering all possible outcomes and assessing each one individually and how one would deal with it, an enlightened researcher puts himself/herself in the place of being able to deal with every eventuality or contingency. None of us can really foretell the future. We may have strong beliefs that may or may not be proven true. Some would call these prejudices.* A more enlightened or charitable view would call these predictions made on the basis of experience and intuition. In any case, we may be right or we may be wrong. We need to plan and be able to reach conclusions even in the unfortunate and sometimes embarrassing situation in which we are proven to be wrong!

9.3 Concluding Remarks

I have listed above what I believe to be the most essential and critical features of experimental design. Perhaps the next most significant issue in design would be the elimination or minimization of experimental bias. We have discussed quite a number of approaches to investigating, quantifying, and ultimately reducing bias in a study. Among the most important techniques considered were the use of uniformity trials to assess both bias and variability, development and adherence to a standard protocol, randomization in various stages of a study, training of investigators and evaluators, and blinding of evaluators as to the treatment group identity of subjects.

The final consideration in designing a study is its efficiency. How many subjects are necessary? Here we discussed the general issue of sample size estimation. Again, exploration and control of sources of

---

* Many statisticians and researchers have preconceived notions about the results of certain studies and incorporate these beliefs in a formal way into data analysis. In the simplest form, such beliefs are incorporated by restricting attention to particular statistical procedures or models. In a more complex way, Bayesian statisticians incorporate their beliefs by defining prior descriptions of the underlying study population (called a prior distribution of the parameters of the model).

variability are also important. The experimenter can consider a variety of possible experimental designs (as discussed in Chapter 5 and elsewhere) and select the one which is most efficient, considering the various tradeoffs between alternative designs. When there are multiple levels of sampling, one also needs to estimate the appropriate number of replicates at each level.

Box and Tiao (1992) talk about the iterative nature of scientific investigation. One has a preliminary hypothesis (conjecture) that he/she wishes to test. This leads to a designed experiment, the analysis of the resulting data, and the formulation of a tentative model of the underlying biological process. The experimental data and the tentative model lead to another or a refinement of the original hypothesis, so that a new iteration of the process is begun.

I view the role of the statistician and statistical thinking as critical to this process. By working closely with our biomedical colleagues, we can assist throughout the process and improve its likelihood of success and its "rate of convergence." Only by creating such a multidisciplinary team can we hope to draw upon the best talents and diverse ideas that we all have. I hope that by considering and utilizing many of the ideas discussed in this book the reader will better understand the statistical concepts applicable to biomedical research and be able to use them in his/her own work. The result should be better experiments and improved medical research.

References

Agresti, A. (1990). *Categorical Data Analysis*, John Wiley & Sons, New York.

Anderson, V. L. and McLean, R. A. (1974). *Design of Experiments: A Realistic Approach*, Marcel Dekker, New York.

Armitage, P. (1955). "Tests for linear trends in proportions and frequencies," *Biometrics*, 11, 375.

Armitage, P. (1975). *Sequential Medical Trials*, Blackwell, Oxford, U.K.

Bishop, Y. V. V., Fienberg, S. E., and Holland, P. W. (1975). *Discrete Multivariate Analysis*, MIT Press, Cambridge, MA.

Box, G. E. P. (1980). "Sampling and Bayes' inference in scientific modelling and robustness," *J. R. Stat. Soc.*, A, 143, 383.

Box, G. E. P. and Cox, D. R. (1964). "An analysis of transformations" (with discussion), *J. R. Stat. Soc.*, B, 26, 211.

Box, G. E. P. and Tiao, G. C. (1992). *Bayesian Inference in Statistical Analysis*, John Wiley & Sons, New York.

Chiacchierini, R. P. and Seidman, M. M. (1991). "Perspectives on clinical studies for medical device submissions," in: *Premarket Approval (PMA) Supplement*, HHS Publication FDA 91-4245, Center for Devices and Radiological Health, Food and Drug Administration, Rockville, MD.

Cochran, W. G. (1954). "Some methods for strengthening the common χ^2 test," *Biometrics*, 10, 417.

Cochran, W. G. and Cox, G. M. (1957). *Experimental Designs*, 2nd ed., John Wiley & Sons, New York.

Conover, W. J. and Inman, R. L. "Rank transformations as a bridge between parametric and nonparametric statistics," *Am. Stat.*, 35, 124.

Cornell, J. A. (1981). *Experiments with Mixtures: Designs, Models and the Analysis of Mixture Data*, John Wiley & Sons, New York.

D'Agostino, R. B. and Stevens, M. A. (1986). *Goodness-of-Fit Techniques*, Marcel Dekker, New York.

DeMets, D. L. (1987). "Practical aspects in data monitoring: a brief review," *Stat. Med.*, 6, 753.

Draper, N. R. and Smith, H. (1981). *Applied Regression Analysis*, 2nd ed., John Wiley & Sons, New York.

Dunn, O. J. (1964). "Multiple comparisons using rank sums," *Technometrics*, 6, 241.

Dunnett, C. W. (1955). "A multiple comparison procedure for comparing several treatments with a control," *J. Am. Stat. Assoc.*, 50, 1096.

Dunnett, C. W. (1964). "New tables for multiple comparisons with a control," *Biometrics*, 20, 482.

Enas, G. G. and Zerbe, R. L. (1993). "A paradigm for interim analyses in controlled clinical trials," *J. Clin. Res. Drug Dev.*, 7, 193.

Fisher, R. A. (1932). *Statistical Methods for Research Workers*, 4th ed., Oliver and Boyd, London.

Fisher, R. A. and Yates, F. (1963). *Statistical Tables for Biological, Agricultural and Medical Research*, 6th ed., Oliver and Boyd, Edinburgh.

Fleiss, J. L. (1981). *Statistical Methods for Rates and Proportions*, John Wiley & Sons, New York.

Fleming, T. R. (1993). "Data monitoring committees and capturing relevant information of high quality," *Stat. Med.*, 12, 565.

Fleming, T. R., Harrington, D. P., and O'Brien, P. C. (1984). "Designs for group sequential tests," *Control. Clin. Trials*, 5, 348.

Geller, N. L. and Pocock, S. J. (1987). "Interim analyses in randomized clinical trials: ramification and guidelines for practitioners," *Biometrics*, 43, 213.

Geller, N. L. and Pocock, S. J. (1988). "Design and analysis of clinical trials with group sequential stopping rules," in *Biopharmaceutical Statistics for Drug Development*, Peace, K. E., Ed., Marcel Dekker, New York, chap. 11.

Guenther, W. C. (1965). *Concepts of Statistical Inference*, McGraw-Hill, New York.

Haybittle, J. L. (1971). "Repeated assessment of results in clinical trials in cancer treatment," *Br. J. Radiol.*, 44, 793.

Hedges, L. V. and Olkin, I. (1985). *Statistical Methods for Meta-Analysis*, Academic Press, New York.

Hollander, M. and Wolfe, D. A. (1973). *Nonparametric Statistical Methods*, John Wiley & Sons, New York.

Hosmer, D. W. and Lemeshow, S. (1989). *Applied Logistic Regression*, John Wiley & Sons, New York.

Jennison, C. and Turnbull, B. W. (1990). "Statistical approaches to interim monitoring of medical trials: A review and commentary," *Stat. Sci.*, 5, 299.

John, P. W. M. (1970). "A fraction of the 4^3 factorial design for estimating main effects," *Biometrics*, 26, 157.

Jonckheere, A. R. (1954). "A distribution free k-sample test against ordered alternatives," *Biometrika*, 41, 133.

Kabasakalian, P., Cannon, G., and Pinchuk, G. (1969). "Fractional factorial experimental design study of benzocaine in throat lozenges," *J. Pharm. Sci.*, 58, 45.

Kendall, M. G. and Stuart, A. (1967). *The Advanced Theory of Statistics*, Vol. 2, 2nd ed., Hafner Publishing Co., New York.

Kupper, L. L. and Hafner, K. B. (1989). "How appropriate are popular sample size formulas?", *Am. Stat.*, 43, 101.

Lawless, J. F. (1982). *Statistical Methods and Models for Lifetime Data*, John Wiley & Sons, New York.

Makuch, R. W. and Johnson, M. (1990). "Active control equivalence studies: planning and interpretation," in *Statistical Issues in Drug Research and Development*, Peace, K. E., Ed., Marcel Dekker, New York.

Mantel, N. and Haenszel, W. (1959). "Statistical aspects of the analysis of data from retrospective studies of disease," *J. Nat. Cancer Inst.*, 22, 719.

Metzler, C. M. (1974). "Bioavailability — a problem in equivalence," *Biometrics*, 30, 309.

Miller, R. G. (1981). *Simultaneous Statistical Inference*, 2nd ed., Springer-Verlag, New York.

National Research Council (1992). *Combining Information: Statistical Issues and Opportunities for Research*, National Academy Press, Washington, D. C.

O'Brien, P. C. and Fleming, T. R. (1979). "A multiple testing procedure for clinical trials," *Biometrics*, 35, 549.

Peto, R., Pike, M. C., Armitage, P., Breslow, N. E., Cox, D. R., Howard, S. V., Mantel, N., McPherson, K., Peto, J., and Smith, P. G. (1976). "Design and analysis of randomized clinical trials requiring prolonged observation of each patient. Part I. Introduction and design," *Br. J. Cancer*, 34, 585.

Phillips, J. A., Cairns, V., and Koch, G. G. (1992). "The analysis of a multiple-dose, combination-drug clinical trial using response surface methodology," *J. Pharm. Stat.*, 2, 1992.

Pocock, S. J. (1977). "Group sequential methods in the design and analysis of clinical trials," *Biometrika*, 64, 191.

Quade, D. (1966). "On analysis of variance for the k-sample problem," *Ann. Math. Stat.*, 37, 1747.

Quintavalla, S. and Parolari, G. (1993). "Effects of temperature, a_w and pH on the growth of *Bacillus* cells and spores; a Response Surface Methodology study," *Int. J. Food Microbiol.*, 19, 207.

Rockhold, F. W. and Enas, G. G. (1993). "Data monitoring and interim analyses in the pharmaceutical industry: ethical and logistical considerations," *Stat. Med.*, 12, 471.

Roush, W. B., Petersen, R. G., and Arscott, G. H. (1979). "An application of Response Surface Methodology to research in poultry nutrition," *Poult. Sci.*, 58, 1504.

Scheffé, H. (1958). "Experiments with mixtures," *J. R. Stat. Soc.*, B, 20, 344.

Schuirmann, D. J. (1987). "A comparison of the two one-sided tests procedure and the power approach for assessing the equivalence of average bioavailability," *J. Pharmacokinet. Biopharm.*, 15, 657.

Selwyn, M. R. (1995). "The use of trend tests to determine a no observable effect level (NOEL) in animal safety studies," *J. Am. Coll. Toxicol.*, 14, 158.

Selwyn, M. R. , Dempster, A. P., and Hall, N. R. (1981). "A Bayesian approach to bioequivalence for the 2 × 2 changeover design," *Biometrics*, 37, 11.

Selwyn, M. R. and Gaccione, P. P. (1993). "Assessing normality and equivariance in small multigroup samples," *Commun. Stat. Sim.*, 22, 61.

Selwyn, M. R. and Rentko, V. T. (1995). "A modified double triangular test and its application to a veterinary clinical study," presented at the Joint Statistical Meetings, Orlando, FL.

Shih, W. J. (1992). "Sample size reestimation in clinical trials," in *Biopharmaceutical Sequential Statistical Applications*, Peace, K. E., Ed., Marcel Dekker, New York.

Terpstra, T. J. (1952). "The asymptotic normality and consistency of Kendall's test against trend, when ties are present in one ranking," *Indag. Math.*, 14, 327.

Tukey, J. W. (1953). "The problem of multiple comparisons," unpublished manuscript.

Wald, A. (1947). *Sequential Analysis*, John Wiley & Sons, New York.

Wallis, W. A. (1980). "The statistical research group, 1942-1945," *J. Am. Stat. Assoc.*, 75, 320.

Walters, L. (1993). "Data monitoring committees: the moral case for maximum feasible independence," *Stat. Med.*, 12, 575.

Welch, B. L. (1947). "The generalization of 'Student's' problem when several different population variances are involved," *Biometrika*, 34, 28.

Westlake, W. J. (1972). "Use of confidence intervals in analysis of comparative bioavailability trials," *J. Pharm. Sci.*, 61, 1340.

Westlake, W. J. (1973). "The design and analysis of comparative blood-level trials," in *Dosage Form Design and Bioavailability*, Swarbrick, J., Ed., Lea and Febiger, Philadelphia.

Whorton, E. B., Jr., Bee, D. E., and Kilian, D. J. (1979). "Variations in the proportion of abnormal cells and required sample sizes for human cytogenetic studies," *Mut. Res.*, 64, 79.

Whitehead, J. (1983). *The Design and Analysis of Sequential Clinical Trials*, Ellis Horwood Ltd., Chichester, U.K.

Whitehead, J. (1992). *The Design and Analysis of Sequential Clinical Trials*, 2nd ed., Ellis Horwood Ltd., Chichester, U.K.

Wittes, J. (1993). "Behind closed doors: the data monitoring board in randomized clinical trials," *Stat. Med.*, 12, 419.

Yates, F. (1935). "Complex experiments," *J. R. Stat. Soc. Suppl. 2*, 181.

Zelen, M. (1969). "Play-the-winner rule and the controlled clinical trial," *J. Am. Stat. Assoc.*, 64, 134.

Appendix A

Glossary of Statistical Terms

accuracy: the degree to which a measurement process is free of bias.

alternative hypothesis: a hypothesis which is presumed to hold if the null hypothesis does not; the alternative hypothesis is necessary in deciding upon the direction of the test (two-sided or one-sided in a specific direction) and in estimating sample sizes.

bias: the long-run difference between the average of a measurement process and its true value.

blinding: the condition under which individuals are uninformed as to the treatment condition of experimental subjects.

block: a set of units which are expected to respond similarly as a result of treatment.

completely randomized design: a design in which each experimental unit or subject is randomized to a single experimental condition or set of treatments.

confidence coefficient: the probability, prior to the study, that a confidence interval will contain the population parameter under consideration.

confidence interval: a random interval depending on study data which, prior to the study, will contain the true parameter with a prespecified probability.

confidence statement: a statement defining the confidence interval, the parameter under consideration, and the degree of confidence.

confounding: the degree to which the effects on an experimental outcome of two or more factors cannot be separately assessed.

continuous data: data which, in theory, could take on any possible value in a given range, even an infinite range; in practice, the number and range of distinct values should be great enough so that the data would not be considered discrete.

critical value(s): the cutoff or decision value(s) in hypothesis testing which separates the acceptance and rejection regions of a test.

crossover design: a special case of the randomized block design in which subjects constitute the blocks and in which treatments are administered sequentially in time; the order of treatments is usually balanced across all subjects.

degree of confidence: see confidence coefficient.

dichotomous data: data which can only take on two distinct values (e.g., success or failure).

double-blinded study: a study in which both the experimental subjects and the investigators are uninformed as to subjects' treatment status.

estimation: an inferential process where the desired result is a single value (point estimate) estimating a population quantity (parameter).

experimental unit: the smallest unit to which different treatments or experimenal conditions can be applied.

false negative: the error of accepting the null hypothesis when it is false.

false positive: the error of rejecting the null hypothesis when it is true.

fractional factorial design: a design in which only a systematic subset (fraction) of all possible treatment combinations of a full factorial design are included; this type of design purposely confounds main effects and low order interactions with high order interactions.

hypothesis testing: a formal statistical procedure where one tests a particular conjecture (hypothesis) on the basis of experimental data.

level of significance: the allowable rate of false positives set prior to analysis of data.

null hypothesis: a hypothesis indicating "no difference" which will either be accepted or rejected as a result of a statistical test.

one-sided test: a statistical test for which the rejection (critical) region consists of either very large or very small values of the test statistic, but not both.

p-value: the probability of obtaining a test statistic as extreme as or more extreme than the observed one; small p-values are unlikely when the null hypothesis holds.

parameter: a population quantity of interest.

pilot study: a preliminary study performed to gain initial information to be used in planning a subsequent, definitive study; pilot studies are used to refine experimental procedures and provide information on sources of bias and variability.

population: the collection of all subjects or units about which inference is desired.

power: the probability of rejecting the null hypothesis when it is false (and hence some specific alternative holds).

precision: the degree to which a measurement process is limited in terms of its variability about a central value.

process optimization: a combination of experimental design and analysis techniques to optimize the response in a process, but without relying on the specific polynomial modeling as in RSM.

protocol: a document describing the plan for a study; protocols typically contain information on the rationale for performing the study, the study objectives, experimental procedures to be followed, sample sizes and their justification, and the statistical analyses to be performed.

randomization: a well-defined stochastic law for assigning experimental units to differing treatment conditions; randomization may be applied elsewhere within an experiment.

randomized block design: a design in which all treatments are applied within each block and treatments are compared within the blocks.

response surface methodology (RSM): the combination of design and analysis techniques to locate an optimum response within the experimental region; the response is modeled using low degree polynomials with cross-factor terms.

sample (random sample): the collection of subjects or experimental units actually included in a study; a sample from a population is considered a random sample when all subjects have an equal chance of inclusion in the sample and when subjects in the population can be considered as independent units.

sensitivity analysis: the process of varying assumptions and/or specifications in order to assess their effects on a result of interest.

single-blinded study: a study in which the experimental subjects are uninformed as to their treatment status.

statistic: a function of the observed data.

subsampling: the situation in which measurements are taken at several nested levels; the highest level is called the primary sampling unit; the next level is called the secondary sampling unit, etc.

test statistic: a statistic used in hypothesis testing; extreme values of the test statistic are unlikely under the null hypothesis.

two-sided test: a statistical test for which the rejection (critical) region consists of both very large and very small values of the test statistic.

variability: the random fluctuation of a measurement process about its central value.

Appendix B

Formulas for Sample Size Estimation

As discussed in Chapter 4, the appropriate formula for sample size estimation depends on the statistical context. Accordingly, we present formulas suitable in the context of point estimation in Section B.1, in the context of interval estimation in Section B.2, and in the context of hypothesis testing in Section B.3.

B.1 Sample Size Formulas for Point Estimation

In the context of point estimation, a standard formula for sample size estimation is given by

$$N = Z^2(\sigma/\delta)^2, \qquad (\text{B-1})$$

where Z is an upper percentile point from the standard normal distribution, σ is the assumed standard deviation, and δ is the required upper bound on the tolerance for the estimator. Formula B-1 applies when the estimator (i.e., the statistic being used) is normally distributed or can be considered approximately normally distributed based on asymptotic (large sample) theory.

For example, suppose that $\sigma = 1$ and that we want our estimate to be within ± 0.25 units (i.e., $\delta = 0.25$) and we want this relationship to hold with 95% confidence. To do this, we set $Z = 1.96$ (the 97.5th percentile from the standard normal), and thus

$$N = 1.96^2(1/0.25)^2 = 61.5.$$

However, because this is a sample size formula, we round up to $N = 62$. The confidence statement applying in B-1 holds only when σ is known. Thus, if σ has been estimated from a small sample, one might consider a range of possible σ values and perform sample size estimation for this range.

B.2   Sample Size Formulas for Interval Estimation

If we are again considering sample size estimation for a single population and wish to construct a confidence interval for a population parameter under the assumption that the statistic being utilized is normally distributed with known variance, then formula B-1 is also appropriate in this situation. As indicated by Kupper and Hafner (1989), however, this formula will typically underestimate the required sample size when the variance is unknown. The reason for this is that, with variance unknown, the sampling distribution of the statistic is not normally distributed because the variance of the distribution must be estimated based on sample data and formula B-1 does not take into account the stochastic nature of the sample variance. For example, if we are constructing a confidence interval for the mean of a normal population, then the sampling distribution of the sample mean will be *t*-distributed, not normal. To overcome this shortcoming, Kupper and Hafner provide a table (their Table 1) relating sample sizes calculated via B-1 to those appropriate with unknown variance (see also Guenther, 1965).

In constructing a confidence interval on the difference between the means of two populations, each assumed to follow a normal distribution with the same variance in each population, a standard sample size formula is given by

$$N = 2Z^2(\sigma / \delta)^2, \tag{B-2}$$

where N is the sample size in each group, Z is as above, and δ is the required bound on the difference between means. As can easily be seen, this is identical to B-1 except for the multiplicative factor 2. For the same reasons given above, this formula is appropriate only with known variance. When the common variance is unknown and must be estimated from the pooled sample variance from the two samples, the sampling distribution of the difference between sample means is again *t*-distributed, and B-2 will lead to an underestimate of the sample size. In their Table 2, Kupper and Hafner (1989) provide correction factors for the confidence interval problem comparing two normal means.

Another common confidence interval problem concerns the difference in response rates for two populations. In studies with an active control (Section 5.5.2) and in which the primary outcome is the comparison of response rates, we may wish to show (with a high degree of confidence) that the difference in rates between the test and (active) control populations is small. We do this under the assumption that the two rates are actually the same.

Makuch and Johnson (1990) provide the following sample size formula (per group) for this situation:

$$N = 2P(1 \pm P)(Z_1 + Z_2)^2 / \delta^2. \qquad (B\text{-}3)$$

Here P is the common response rate, Z_1 and Z_2 are upper percentile points of the standard normal distribution whose use is defined below, and δ is the required bound on the difference in response rates.

There are two normal percentiles in formula B-1. The first, Z_1, indicates the degree of confidence. For example, 1.96 would be used with a two-sided 95% confidence interval. The value 1.645 would be used for a one-sided 95% confidence limit, i.e., an upper or lower bound on the difference between rates. The second percentile, Z_2, provides the probability that the selected bound will obtain.

As an illustration, consider the sample sizes displayed in Table 4.1. With $P = 0.8$, $\delta = .1$, $Z_1 = 1.645$, and $Z_2 = 0.842$ (80% probability of bound holding), we obtain

$$N = 2(0.8)(1 \pm 0.8)(1.645 + 0.842)^2 / (0.1)^2 = 197.9.$$

We therefore round up to 198, the value given in Table 4.1.

B.3 Sample Size Formulas for Hypothesis Testing

Sample size formulas used in the context of hypothesis testing are also quite straightforward. For example, suppose that one is interested in testing the null hypothesis that the mean of a normal population with known variance is equal to zero (or some other quantity). The appropriate sample size formula is then given by

$$N = (Z_1 + Z_2)^2 (\sigma / \delta)^2. \qquad (B\text{-}4)$$

As above, Z_1 and Z_2 are upper percentiles from the standard normal distribution, σ denotes the population standard deviation, and δ is the population mean under the alternative hypothesis. The first percentile, Z_1, is determined based on the level (false positive rate) of the test. The second percentile, Z_2 is related to the power of the test. For example, if one is performing a one-sided test at the 0.05 level, then $Z_1 = 1.645$. If the test is two-sided and at the 0.05 level, the value used is 1.96. Similarly, if one desires that the test will have 80% power when the alternative holds (i.e., the true mean is δ), then Z_2 takes on the value

0.842. As an example, when $\delta = \sigma$, a two-sided 5% level test with 80% power requires $N = (1.96 + 0.842)^2 (1)^2 = 7.9$ (i.e., 8) subjects.

If one is comparing the means of two normal populations with known and equal variances, then the sample size formula (N per group) to test the hypothesis of equality of means is given by

$$N = 2(Z_1 + Z_2)^2(\sigma/\delta)^2. \tag{B-5}$$

Here, Z_1, Z_2, and σ are as given above, and δ is the postulated difference between the two population means under the alternative hypothesis. Kupper and Hafner (1989) indicate that formulas B-4 and B-5 are satisfactory when the underlying variance is unknown, except in exceedingly small samples.

Sample size formulas along the lines of B-4 and B-5 are appropriate in other hypothesis testing situations. For example, suppose that one is considering an experiment with K (normal) populations and one wishes to test a hypothesis about a linear combination of the means. In the dose-response case with four groups, equally spaced and ordered coefficients such as -3, -1, 1, and 3 could be used to indicate an overall trend in the population means. If we were to use a test based on this linear contrast, what sample size per group would be required? If we let c_1, c_2, c_3, and c_4 denote the above coefficients, define our test statistic as

$$Z = \frac{\sum_{i=1}^{4} c_i \bar{x}_i}{\sigma \sqrt{\sum_{i=1}^{4} c_i^2 / N}}, \tag{B-6}$$

then the required sample size per group is given by

$$N = \left(\sum_{i=1}^{4} c_i^2\right)(Z_1 + Z_2)^2(\sigma/\delta)^2, \tag{B-7}$$

where $\delta = \sum_{i=1}^{4} c_i \mu_i$ is the true value of the linear contrast. It can easily be observed, for $K = 2$ instead of 4 and with coefficients $c_1 = 1$ and $c_2 = -1$, formula B-7 reduces to formula B-5 and δ reduces to

$(1)\mu_1 + (-1)\mu_2 = \mu_1 - \mu_2$, just the difference between the population means.

Another common hypothesis testing situation is the comparison of two populations with regard to a dichotomous outcome. Suppose that P_1 is the response rate in population one and that P_2 is the response rate in population two. An approximate sample size formula for comparing these rates (Fleiss, 1981) is given by

$$N = \frac{\left(Z_1\sqrt{2\bar{P}\,\bar{Q}} + Z_2\sqrt{P_1 Q_1 + P_2 Q_2}\right)^2}{(P_1 \pm P_2)^2} \tag{B-8}$$

where

$$Q_1 = 1 \pm P_1, Q_2 = 1 \pm P_2, \bar{P} = \frac{1}{2}(P_1 + P_2), \text{ and } \bar{Q} = 1 \pm \bar{P}.$$

For example, with $P_1 = 0.8$ and $P_2 = 0.5$, a two-sided 5% level test with 80% power would require each group to be of size

$$N = \frac{\left(1.96\sqrt{2(0.65)0.35} + 0.842\sqrt{0.8(0.2) + 0.5(0.5)}\right)^2}{(0.8 \pm 0.5)^2} = 38.5.$$

However, as indicated by Fleiss (1981), this formula tends to underestimate the true N required. He therefore proposes a further correction

$$N^* = \frac{N}{4}\left[1 + \sqrt{1 + \frac{4}{N|P_1 \pm P_2|}}\right]^2 \tag{B-9}$$

Application of Formula B-9 to the numerical problem above gives $N^* = 45$ per group. (See Section 6.3 for an example in which this calculated sample size was used.)

Index

–2–

2^K design, 56

–A–

accuracy, 17, 147
adjusted t-test, 129
alias, 104, 108
alpha level. *See* false positive rate
alternative hypothesis, 42, 43, 44,
        49, 84, 87, 129, 147
analysis of covariance, 130
analysis of variance, 98, 121, 124,
        127, 129, 132
ANCOVA. *See* analysis of
        covariance
ANOVA. *See* analysis of variance

–B–

balanced complete block design,
        58
balanced incomplete block
        design, 65
bias, 4, 5, 17–27, 30, 32, 34, 35, 59,
        61, 67, 68, 70, 130, 140,
        147
binary data. *See* dichotomous
        data
bioequivalence, 6, 12, 62, 71, 72
blinding, 5, 19, 24, 34–35, 69, 79,
        147

block, 5, 23, 24, 32, 147
blocking, 5, 23, 32-33, 60, 79, 80

–C–

carry-over effect, 64, 65
central composite design, 110, 111
central composite rotatable
        design, 113
chi-square test, 133, 134
Cochran-Armitage trend test,
        134
Cochran-Mantel-Haenszel test,
        134, 135
completely randomized design,
        5, 12, 28, 54–58, 61, 65, 73,
        98, 103, 147
composite design, 110, 111, 113
confidence coefficient, 39, 50, 147
confidence interval, 38, 40, 42, 50,
        72, 127, 147, 152
confidence limit, 153
confidence statement, 40, 147
confounding, 29, 68, 104–108, 110,
        147
contingency table, 126, 133, 134
continuous data, 38, 46, 119, 120,
        121, 124, 125, 127–133,
        147
controls, 67–72
    active, 70–72
    negative, 68–70
    placebo, 69–70

subjects as own, 68
  untreated, 69
covariate, 29, 130
critical value, 95, 132, 147
crossover study, 6, 62–65, 148

–D–

data analysis, 119–136
data monitoring board,
    99–102
defining contrast, 105–107
degree of confidence. *See*
    confidence coefficient
diagnostic plotting, 122
dichotomous, 87, 120, 134, 155
dichotomous data, 46, 87, 120,
    134, 148
discrete data, 38, 119–123, 133,
    134
distribution-free methods. *See*
    nonparametric methods
dose-response, 72–77
double triangular test, 91
double-blinding, 24, 34, 70
dropouts, 4, 21, 24, 61

–E–

efficacy population, 91
exact permutation trend test,
    134
experimental unit, 5, 26–27, 28,
    47, 48, 148

–F–

factorial design, 54, 55–58, 110,
    129
false negative, 42, 148
false positive, 42, 45, 84, 124, 148
Fisher's exact test, 133
fixed effect, 59
fractional factorial design, 6,
    103–110, 148

–G–

Gaussian distribution. *See* normal
    distribution
group sequential design, 6, 93–99
group sequential test, 85

–H–

historical controls, 67
hypothesis testing, 5, 37, 38, 39,
    41–45, 48, 71, 86, 122, 127,
    148, 153–155

–I–

intent-to-treat population, 91
interaction, 57, 66, 105, 107
interference. *See* interaction
interim analysis, 3, 6, 85, 86, 91,
    94, 95, 98, 99, 100, 102
interval data, 120
interval estimation, 5, 37, 38, 77,
    152–153

–J–

Jonckheere's test, 131, 132

–K–

Kruskal-Wallis test, 131, 132

–L–

Latin Square design, 63
level of significance. *See* false
    positive rate
linear contrast, 154
logistic regression, 134

–M–

matched pair design, 60
meta analysis, 119, 135–136

monotonic dose-response, 73

multicenter study, 13, 29, 78–80, 85

multiple comparison methods, 129, 132

multiplicity adjustments, 123–124

–N–

nominal data, 120–121

noncentral composite design, 112

nonparametric methods, 121, 123, 126, 127, 130–133

normal distribution, 95, 121, 126, 130, 132, 151, 152

null hypothesis, 39, 41, 42, 43, 44, 84, 87, 90, 98, 122, 129, 130, 148

–O–

objectives, 1, 3, 4, 6, 7, 9–11, 13, 18, 37, 38, 54, 67, 72, 77, 78, 119

one-sided test, 43, 44, 148

one-tailed test. *See* one-sided test

one-way layout, 54, 55, 124

ordinal data, 99, 120, 121, 133

–P–

paired t-test, 131

parameter, 38, 39, 40, 50, 127, 148

parametric methods, 121, 126, 127–130

pilot studies, 5, 25, 37, 39, 45–48, 148

pivoting, 127

plan. *See* randomization plan

point estimation, 37, 38, 77

population, 9, 19, 38, 39, 40, 48, 50, 59, 68, 85, 122, 148

power, 42, 45, 49, 50, 92, 130, 149

power transformation, 126

precision, 17, 33, 37, 39, 48, 49, 73, 113, 149

process optimization, 6, 103, 114–117, 149

protocol, 4, 6, 10, 11, 12–13, 22, 46, 86, 91, 99, 102, 139, 140, 149

p-value, 44, 136, 148

–R–

random effect, 59, 135–136

random sample, 19, 149

randomization, 4, 5, 11, 19, 24, 27–29, 33, 58, 80, 87, 140, 149

randomized block design, 6, 12, 24, 58–62, 80, 149

ratio data, 120

repeated measures design, 66

reproducibility, 20

response surface methodology, 6, 103, 110–114, 149

robust methods, 122, 132

rotatability, 113

rotatable design. *See* rotatability

RSM. *See* response surface methodology

–S–

sample, 2, 5, 18, 19, 25, 37–51, 127

sample size, 3, 6, 25, 37–51, 78, 79, 83, 84, 88, 91, 93, 94, 127

sample size estimation, 2, 5, 37–51, 86, 94, 120, 127, 151–152

sensitivity, 20, 122

sensitivity analysis, 37, 46, 48, 49–51, 149

sequential clinical trial, 83–102

sequential design, 3, 6, 86, 87, 91, 93

sequential experimentation. *See* sequential clinical trial
sequential probability ratio test, 84, 89
sham control. *See* placebo control
simplex lattice design, 116
single-blinding, 24, 34, 70
specificity, 20, 38, 73
split plot design, 65–66, 67
SPRT. *See* sequential probability ratio test
standard error, 38, 39
statistic, 38, 87, 127, 149, 151
statistical packages, 124–126
stochastic, 152
stopping rule, 86
stratified design. *See* randomized block design
stratum, 32, 33, 58, 59, 60, 80
subjective measurements, 22, 98
subsampling, 5, 21, 25, 37, 48–49, 66, 149
switch-back design, 65
synergy. *See* interaction

–T–

test statistic, 43, 44, 127, 149
transformation of data, 126

trend test, 124, 130, 134
triangular test, 85, 89, 93
t-test, 128, 129
two period cross-over design, 12, 62
two-sample Wilcoxon test, 132
two-sided test, 44, 45, 50, 92, 132, 149, 155
two-tailed test. *See* two-sided test
Type I error. *See* false positive
Type II error. *See* false negative

–U–

uniformity study, 2, 30–32

–V–

variability, 2, 4, 5, 15, 17–27, 30, 31, 37, 38, 39, 40, 46, 51, 68, 149
vehicle control. *See* placebo control

–W–

washout period, 12, 61, 63
Wilcoxon signed rank test, 130, 131